PORT-CITY INTERPLAY

Transport and Mobility Series

Series Editors: Richard Knowles, University of Salford, UK and Markus Hesse, Université du Luxembourg and on behalf of the Royal Geographical Society (with the Institute of British Geographers) Transport Geography Research Group (TGRG).

The inception of this series marks a major resurgence of geographical research into transport and mobility. Reflecting the dynamic relationships between socio-spatial behaviour and change, it acts as a forum for cutting-edge research into transport and mobility, and for innovative and decisive debates on the formulation and repercussions of transport policy making.

Also in the series

The Geographies of Air Transport
Edited by Andrew R. Goetz and Lucy Budd
ISBN 978 1 4724 2689 5

Innovation in Public Transport Finance
Property Value Capture
Shishir Mathur
ISBN 978 1 4094 6260 6

Hub Cities in the Knowledge Economy
Seaports, Airports, Brainports
Edited by Sven Conventz, Ben Derudder, Alain Thierstein and Frank Witlox
ISBN 978 1 4094 4591 3

Institutional Barriers to Sustainable Transport
Carey Curtis and Nicholas Low
ISBN 978 0 7546 7692 8

Daily Spatial Mobilities
Physical and Virtual
Aharon Kellerman
ISBN 978 1 4094 2362 1

Territorial Implications of High Speed Rail
A Spanish Perspective
Edited by José M. de Ureña
ISBN 978 1 4094 3052 0

Port-City Interplays in China

JAMES JIXIAN WANG

University of Hong Kong, China

LONDON AND NEW YORK

First published 2014 by Ashgate Publishing

Published 2016 by Routledge
2 Park Square, Milton Park, Abingdon, Oxon OX14 4RN
711 Third Avenue, New York, NY 10017, USA

First issued in paperback 2018

Routledge is an imprint of the Taylor & Francis Group, an informa business

British Library Cataloguing in Publication Data
A catalogue record for this book is available from the British Library

The Library of Congress has cataloged the printed edition as follows:

Wang, James Jixian.
Port-city interplays in China / by James Jixian Wang.
   pages cm – (Transport and mobility) Includes bibliographical references and index.
   ISBN 978-1-4724-2689-5 (hardback) 1. Port cities–China. 2. Port
cities–China–Economic conditions. 3. Harbors–China–Economic conditions. 4.
Harbors–Government policy–China. 5. Globalization–Economic aspects–China. I.
Title. II. Series: Transport and mobility series.
   HT169.C6.W3746 2014
   307.760951–dc23

                                                                                 2013045807

ISBN 13: 978-1-138-54665-3 (pbk)
ISBN 13: 978-1-4724-2689-5 (hbk)

Contents

List of Figures

List of Tables

Chapter 1

Introduction:
Coastal Port Cities as Global Interfaces

1.1 Preface

Fifty US cents is all that it takes to ship a pair of brand name sports shoes, priced at 50 dollars, from its manufacturing location in the Pearl River Delta in China to Wal-Mart stores in big cities across the United States. China has become part of the global market economy as a result of the development of the international shipping industry and the reform and opening-up in China in the last three decades. This has changed the world economy thoroughly because China is today a very important link in globalization. This link means not only material exchanges, political conflicts, cultural shocks and access, but also change in regions and cities in China itself that have become involved in the process of globalization. These regions and cities are concentrated in coastal areas, since the actual hubs of foreign trade and foreign exchange channels in the last two decades have been concentrated in coastal areas. This concentration trend is increasing rather than decreasing. As of 2007, foreign trade in coastal provinces and cities accounted for 93 percent of the total foreign trade in China. No matter whether this kind of development mode is successful or not, the significant changes during the last two decades themselves deserve careful examination. As a geographer born and grown up in China and working in Hong Kong, I have the research interest as well as the responsibility to provide explanations for the development and evolution of port-city relationships brought about by globalization from the perspective of transport geography.

Water transportation is an old and inexpensive form of transportation. Port cities are clusters of people and social and economic activities that evolved from places serving as water transportation connection points. Of course, these cities have something important in common. They can achieve economic prosperity through trade with other regions and cities by water. From Quanzhou during the Ming Dynasty to Shanghai and Hong Kong in the twentieth century, from London to New York and Los Angeles, from Singapore in Southeast Asia to Cape Town in South Africa, even to inland harbor cities like Chicago, Manchester, and Wuhan, they all developed their economies in this way. Therefore, in recent years, phrases like "developing cities through ports and vice versa" and "using ports to develop cities and vice versa"[1] have often appeared in various documents, speeches and

---

1   For example, 1) "Party Secretary of Taicang City: 'Developing city through port' and 'hooking up with Shanghai'", A China.com.cn report, August 2003, available at http://www.

reports on port cities' development as a way to describe the interactive relationships between ports and cities in China. In 2002, the Ministry of Communications in China also stated that port-city relationships are one of the top five relationships that need to be handled in China's port industry[2]. However, if we think about this a bit more, we will find that the questions related to relationships between ports and cities are not easy to answer: How can we achieve "developing cities through ports and vice versa"? Can ports "promote the development of cities"[3] 100 percent? Does the development of cities necessarily promote the development of ports? At what levels do ports and cities have interactive relationships?

In fact, these questions exist all over the world. They are the questions that many transportation geographers have been concerned with over the past half century. In China, Huang Shengzhang (1951) may be the first person to have studied port-cities as a whole from the academic perspective. For the study of modern port cities in foreign countries, the early "Anyport Model" (Bird 1963, for details, see Chapter 4) first began to attract relatively large attention of scholars. As shipping technologies develop, international, political and economic relationships and government models change, and as the multimodal transport of logistics appears overseas, the study of port cities has also developed. Bird's study is concentrated on the topology of interaction between ports and cities in the evolvement of British port cities. In the following decade, and largely influenced by development economics, transport geographers studied the evolution of port cities in Asian and African colonies where colonist countries were obtaining natural resources. A typical example is the so-called TMG model (Taaffe, Morrill and Gould 1963) derived from the analysis of individual African cases. Later, Hoyle applied this model to other African countries (Hoyle 1968), and Rimmer conducted a similar analysis for southeastern Asian countries (Rimmer 1967). At that time, port development was relatively slow and port cities experienced orderly, spatial expansion. With the emergence of container transportation in the 1990s, marine transport and ports have witnessed rapid changes. Container ports developed in several phases (Hayuth 1987). The world's major coastal cities began to compete to become the container port hub. The competition has become fiercer with bigger ships and the appearance of the "hub-and-spoke" regional port system. After this, scholars began to concentrate their studies on the hub port and its evolution. The privatization of the port industry in the world and the trend of

---

china.com.cn/chinese/zhuanti/djyzj/391108.htm ; 2) Wang, Cainan, (2000) "Strenthing the Strategy of 'To Thrive the City by Relying on the Port and to Boot the Port by Replying on the City'", Journal of Ningbo University, 13(1), pp. 4–6; 3) Xie, Zhenming (2004), "Thoughts on building Ningbo modern international port", Port Economy (Chinese), Vol 5, p. 12. 4) a report on Yangcheng Evening News on July 31, 2013: "Provincial governor Zhu Xiaodan emphasizing to develop Yangjiang City relying on port and industrial cluster at port".

2   See "Speech Made in National Conference on Port Management" by Zhang Chunxian, December 19th, 2002, Beijing.

3   Ditto.

transnational operations brought about the revolution in global port management. Many countries, including China, allowed transnational port operators to invest and operate container ports in their countries in order to facilitate the integration of their economies into the global economy, global supply and production chains. Large liner companies one after the other began to invest in the port business, and even the land logistics businesses, thus increasing their market shares through vertical integration of the logistics chain and diverting the risks of the shipping industry. These changes shifted academic studies to the regional development of ports (Notteboom 2007). However, for studies on port cities and the port-city relationship, developed countries gradually began to conduct these studies from a post-modern perspective. For example, they paid a lot of attention to various social problems brought on by redevelopment of the waterfront shoreline (Hilling and Hoyle 1984; Hoyle 1989, Pinder and Slack (eds) 2004). This is because globalization led to manufacturing relocation; most once-prosperous ports have entered the late stage age, and only a few large scale regional hubs such as Rotterdam and Antwerp are still concerned with new problems of port-city relationship brought about by expanding global trade.

One major problem concerns the environment. Maritime transport may contribute much less CO_2 than trucking and air transport, in terms of per ton-kilometer of cargo shipped. However, considering the total traffic handled a regional hub port for maritime shipping is still regarded as a major source of congestion on land transport networks as well as a major source of carbon emission for its host city, because of the intensive and direct consumption of fossil-based fuels by the terminal and vessel operations at the harbor. One sustainable solution to this port-city conflict is to place the port further away from the city. However, Hall and Jacob (2012) argued that major ports in developed countries remain urban largely because of the dependency of the port and its cluster of related logistics activities on supporting services that are available only in developed and wealthy cities. Therefore, the mediation of environmental problems becomes a key concern in the policy agenda of the city government, given that the port is no longer a popular sector that generates much employment.

Understanding the port-city dialogue is limited neither to the mutual dependence of the port and the city nor to the sustainability of these so-called port cities. Two dimensions that have been studied in depth are port regionalization and the port-city as a terminal of flows. Port regionalization is a trend or new stage in the concept of the port-city-region, wherein a rescaling of the port function occurs, thus forming a regional system that serves a higher and more diversified demand at a global level (e.g., Notteboom and Roglegue 2006; Notteboom 2007; Hense 2013; Oliver and Slack 2007). This rescaling process is largely initiated and driven by global trade and its major players (e.g., international shipping lines and container terminal operators), and is the research focus of a number of studies. However, the displacement of the port city has complicated not only the relationship between ports and terminal operators but also the interrelationship among the port cities involved. For example, given its role in the intercontinental gateway, Vancouver in Canada must consider its relation with Seattle as a continental corridor for North America. Similarly, Hong Kong,

Shenzhen, and Guangzhou must cooperate to maintain regional competitiveness in the global manufacturing business. In this process of co-petition, each individual port must differentiate its services to attract more shipping lines to call. The local state must also be responsible for facilitating the trade and for nurturing the growth of "related variety" around the port area (Frenken, van Oort, and Verburg 2007) to gain from the added value generated by global supply chains.

The rescaling of port development and its impact on the port city from a maritime perspective reveals several arguments concerning the role of the port city as a terminal for regional, national, and global flows from an urban perspective. These compelling arguments demonstrate the complications of port cities in the contemporary globalized world. According to Hasse (2008; 2012; 2013), cities have been historically understood as market places that provide "regular exchanges of goods" (Weber 1921) as well as services and goods both within and outside the area (Christaller 1933). During the era of industrialization and urbanization, a hierarchical structure of gateway functions was established for local, regional, and interregional gateway cities, in order to handle the different demands of flows. Such a structure is reshaped by the changing land-use brought by suburbanization and other urban processes. However, the structure is also challenged by globalization, which has resulted in the relocation of distribution centers (DCs) away from the port and city center (van Klink 2002), which can be a larger problem when space and land at the port and inner city areas became either too expensive or congested. Such a change is merely a part of the story. Hasse (2008) argued that globalized distribution regimes have redesigned the urban logistics system as well as those systems and networks that serve city-regions, continents, or even the global economy. Therefore, the city is turned into a terminal, in which logistics and freight distribution (where manufacturers are shippers and collective or individual consumers are consignees) proactively reshape cities for the flows.

The interlink between urban logistics/freight distribution and maritime port development in this globalized world comprises the so-called global supply chains (GSC). On land, third party logistics (3PL) firms as the chain connector effectively integrate GSCs to follow the market forces and choices. On the seaside, Notteboom (2007) and Olivier and Slack (2007) posited that major shipping lines act on behalf of the GSCs to decide the port of call, which results in either expending an existing port or eventually building a new port as a gateway to that region. These arguments on maritime port development and urban port-city changes in terms of freight distributions reveal that the role of port cities as terminals or intermediate nodal place in the globalized world is largely defined and shaped by GSCs and/or global production networks (GPNs) (Hendersen et al. 2002; Hesse and Rodrigue 2006). This process of port-city positioning varies locally, and brings diversified global-local confrontations that are locally embedded in economic, environmental and social issues, which may continuously imprint on the city making and port-city interface.

Hall and Hesse (2012) argued that such imprints have two spatial processes, namely, the integration (or disintegration) and rescaling to aid cities in coping with the multi-scale demands from transport and logistics. Meanwhile, Hall and Jacobs (2012) suggested that the evolutional economic geography (EEG) perspective can be used to further understand these processes as well as their related causes and consequences. The EEG focuses on the firm or organization at the micro-level. Its unit of analysis consists of the routines and the processes of co-evolution followed by the firms in a space with institutional arrangements, resulting from the interaction-based learning and innovation. Hall and Jacobs stated, "An EEG perspective on why ports are still urban would insist on asking why port actors in urban spaces have an advantage over non-urban port decision-makers. In general terms, the answer is that knowledge, innovation and decision-making capacities remain primarily urban. Global maritime decision-makers, that is, the location of the headquarters 35 of the largest terminal operating companies and shipping lines, are concentrated in only a few cities". (Hall and Jacobs 2012, p. 202). From the same EEG perspective, Jacobs and Notteboom (2011) analyzed the practices or routines of terminal operators in regional seaport competition in Europe.

The literature reviewed above intensively covers the global trends and local issues faced by numerous port cities, including those in China. The views expressed by Hall and Jacobs on the port-city interface (2012) from the EEG perspective and the theoretical angle, as well as the analytical results of the case studies presented by Hasse (2008; 2013) shed light on the new phenomena and mechanisms of the geographical dynamics brought about by global logistics and maritime transport into the port cities.

However, given the substantial differences in the overall institutional setting between China and the developed countries in the West, we argue that a different approach is needed to analyze the dynamics of the port-city interplays in China. The primary concern in properly applying a theory or analytical framework to local dynamics is the fitness of the theory or framework to the stage, trajectory, and institutions of the specific city, region, or country in question. We summarize the three major characteristics of China, which require special treatment in the application of theory, methodology, and approach.

First, China should be studied as an "extractive political and inclusive economic system", as suggested by the institutionalist theory of Acemoglu and Robinson (2012). This theory is based on the conviction that institutions, rather than geographic or cultural factors, explain the vast differences in economics, politics, and social organizations. Political and economic institutions develop along a continuum; however, they can be roughly divided into two categories: inclusive and extractive. The division between political and economic institutions, on the one hand, and inclusive and extractive, on the other hand, create four types of institutional configurations as follows: 1) Inclusive political and Inclusive economic, 2) Extractive political and Inclusive economic, 3) Inclusive political and Extractive economic, and 4) Extractive political and Extractive economic. Since the beginning of the economic reforms implemented by Deng Xiaoping and his allies in 1978, China experienced a drastic change from being a country with political extractive and economic extractive institutions to one with political extractive institutions with economic inclusive institutions. Before Deng and the reformers rose in power, China was under a single party authoritarian system with extractive political institutions that were simultaneously highly extractive economic institutions. The economic reforms they implemented partly reinstalled the market economy with inclusive economic institutions. However, politically speaking, the country remained under the tight control of a single authority through the power of the state. The identification of the 14 coastal port cities as "open cities" is considered a case that illustrates how the extractive political institution turned coastal economy from extractive to inclusive by giving export-oriented processing industries the ability to attract foreign investment (Chapter 15, p. 467).

This special configuration of political and economic institutions was evident through the city-to-city competition during the past three decades of rapid economic growth. A top-down performance management system was implemented from the Central Government to the county governments, where each level of the state was required to compete with their counterparts in terms of GDP generation. Therefore, the governments of the port cities had every reason to promote the ports *within* their jurisdiction to maximize port-related GDP-generating activities. Here, the definition of city is clearly jurisdictional rather than functional, based on the analysis of the port-city interface by Hall and Jacobs (2012). The port city governments, instead of the enterprises, now play the dominant role in determining the future of both port and city. This is because the time and place in making economic institutions more inclusive and market oriented are basically decided through extractive political institutions. In other words, understanding the evolution of *the routines of the states* rather than *the routines of the firms* is the key in understanding the port-city dynamics in China.

Second, the development trajectory of China is another major characteristic that differentiates it from its Western counterparts. Unlike most capitalist countries that have been part of the global market for a long time, China only opened up to the world in 1978. The 30-year isolation from 1949 to 1978 created a huge gap in the market economy between the coastal cities and similar cities in terms of port infrastructure. For example, the first container terminal in Mainland China opened only in 1992, 20 years after those of Singapore and Hong Kong. The technological backwardness was also generally evident in the port draft, because China completely lacked global liner services for three decades. The enlargement of vessels in the maritime shipping market did not have any impact on port upgrading before 1980. Another evident gap is the overall level of containerization. China did not use internationally standardized (ISO) containers until the early 1990s, because back then, the country had little international direct trade through maritime connections. As discussed in latter parts of this book, the use of ISO containers remains lacking because most parts of China, except for coastal cities and regions involved in global trade, dominantly rely on conventional methods of freight transport such as break-bulk trucking and railways. Such a development gap has had a profound impact on the country. For example, the unpopularity of ISO containers (20-foot equivalent unit or TEU and 40-foot equivalent unit or FEU) makes the shipment of containerized cargo to or from the inland regions extremely expensive. Therefore, the consolidation and deconsolidation of container cargo are always located close to port areas in order to facilitate movements from conventional highway transport or shuttle trucking service to the conventional rail freight system.

Third, the port-city dynamics in China is considerably characterized by the developing stage of economy with a strong but unbalanced trade in the era of globalization. All developed countries underwent the so-called "developing stage", but few experienced the sudden change from a closed economy to an open one that immediately faced globalization. Therefore, achieving rapid economic growth by taking advantage of joining the global market results in some trade-offs, such as the low priority given to environment and social equality. This is because China has largely become an economic-growth-driven country where the city-to-city competition is measured by total GDP. Unfortunately, the deteriorating environment and enlarged social gap are overshadowed by the overwhelming economic growth, which is reflected in the extremely rapid port development in the country. Within three decades, the international container throughput of China increased from zero in 1980 to 129.6 million TEUs in 2010, which equals that of the sum of the European Union and the United States of America (Figure 1.1).

Given the three characteristics discussed above, several questions arise. How does the positive relationship between extractive political institutions and inclusive economic institutions in developing port cities lead to the trade dependent economy of China? Alternatively, how does the state – as both referee and player – act in the port-city growing game in response to the market forces represented by various actors (e.g., international terminal operators) in order to facilitate the rapid trade-dependent growth of the country? What evidence or process is available to indicate that a state-led development of ports and port cities may result in different consequences from those found in the market economies? What are the differences in the adjustment processes and contents of different cities when forming or transforming a conventional port to an international articulation place?

No systematic answers are currently available for these questions. One of the reasons is the uniqueness of the macro configuration of socio-economic and political institutions as well as the development stage and trajectory. The participants in individual Chinese port city building and management efforts do pay attention to macro and overall comparisons. Furthermore, only a small number of people from the academe are given the opportunity to participate in development of multiple port-city relationship programs in China. Therefore, the academics are unable to relate the evolution to relevant existing theories and are unable to conduct an overall study. Fortunately, I belong to the small group of people who are given the opportunity to participate in the planning of numerous port cities in the country, and I feel obligated to share the knowledge.

Given the unique trajectory of the port city development and the different characteristics discussed above, an analytical framework that is tailor-made for the case of China must be proposed, as discussed in Chapter 2. This piece of work aims to contribute to the further understanding of port-city dynamics in the

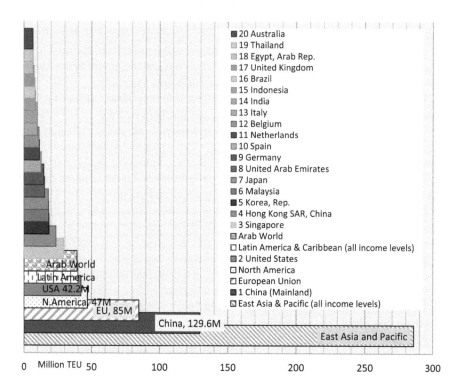

Figure 1.1   Coastal Container Terminal Throughputs: World Major Regions and Top 20 Countries/Regions (2010)

developing world. Such contribution may be complementary in enriching our overall knowledge, particularly how port-city dynamics can be comprehended as a nodal interface or an articulation space, which is shaped by a continuous global-local dialogue of flows.

1.3   Structure of this Book

In China, the development of most coastal cities by mainly taking advantage of their trade interface resources and urban waterfront redevelopment is just beginning. Given this fact, this book focuses on the interactive relationship between ports and cities in China against the background of globalization. This focus determines this book's structure. In Chapter 2, some necessary background knowledge and important definitions need to be dealt with first. It is followed by an analysis framework of the port-city relationship, and then an introduction to major driving factors and policies for port development in coastal cities in China. Based on the framework put forward in Chapter 2, from Chapter 3 to Chapter 6, in-depth discussions about four aspects of port-city relationships in China are set forth: economic relationships, geological relationships, external network relationships and port-city governance relationships. Chapter 7 examines port cities in the regional context. It discusses and illustrates how external factors affect the development of individual port cities according to special conditions in different coastal areas in China. Chapter 8 reviews and comments on the national policies and institutional systems that lead to the new geographical formation of coastal regions in China. Chapter 9 summarizes the analysis of port-city relationships in the previous chapters. At the same time, the chapter further discusses possible macro results and effects if coastal port cities continue to develop following the current trend in China.

Chapter 2
An Analytical Framework for Port-City Relations

2.1 Three Sets of Basic Concepts

Before conducting a comprehensive analysis of port-city relationships, several basic concepts related to ports should be established.

The first set of basic concepts is the definitions of ports and relevant organizations. The first Port Law promulgated in China in 2003 did not define what a port is. UNCTAD categorized ports into four generations (see Table 2.1).

First generation ports refer to the interfaces formed between marine transportation and land transportation before 1960. At that time, ports mainly handled break-bulk and bulk cargoes. Different kinds of shipping sectors were independent from each other. The second generation ports took place from 1960 to 1980. At that time, as the global economy developed, ports gradually became the places where cargo was loaded and unloaded and where services were provided to the industrial and commercial sectors. After 1980, with the development of, the container industry and multimodal transportation, the third generation of ports was born, a generation which gradually became hubs of the international production and distribution networks. Port management here was more proactive. The fourth generation was born after 1990 when port information and operating procedures were standardized, facilities and equipment was automated, operations and management were information-based and globalized and the transnational operation of large port transport enterprises began to appear.

The progress of the "generations" was caused mainly by technological advancement and global economic integration. However, as Beresford and others (2004) have indicated, this port classification is not always clear, nor did ports necessarily develop according to the sequence of generations. In reality, ports from different generations may co-exist together or be distributed among a few ports or port districts within a city or region. On this regard, Bichou and Gray (2005) provide excellent port taxonomy. They argue that as the port role today exceeds the simple function of services to ships and cargo, it is the approaches researchers carry that determines which decisive factors, missions, assets and facilities, functions, and institutions are important and critical. For example, from a geographical and spatial angle, a port-city approach treats the mission of port is the major subject of study, while an approach to analyzing waterfront estate focuses on the assets and facilities of port. In other words, an approach of research towards port-city dynamics from geographical perspective is more macro than

Table 2.1 Worldwide Models of Four Generations of Port

Generation Item Period	The 1st Before 1960s	The 2nd After 1960s	The 3rd After 1980s	The 4th Since 1990s
Main commodities	General cargo	General cargo and bulk cargo	bulk cargo and united cargo (larger ships and bigger capacity)	Cargoes are more united
Concept and strategy for port development	The concept was conservative, ports were regarded only as changing spots for different transportation methods	Dedicated to development and became industrial and commercial hubs	Commercial operation, dedicated to becoming the connection points of multimodal transportation and modern logistics centers	More automated than the third generation
Business priority	1）Cargo interface between ships and banks	1 and 2）Increased cargo transport and industrial activities	1, 2 and 3）Logistics and information flow smoothly and were fully utilized	More standardized and information-based than the third generation
Characteristics of the port organization	Business activities in the port were independent of each other. Ports were not connected closely to the outside	Close relationships among ports and users. Business activities were loosely tied to each other, but any port-city relationship was still temporary.	All connections were intensified and unified. Privatization was put on the agenda	Port organization is more globalized and port area is more ecological
Production characteristics	Only goods with little added value flow	Besides cargo loading and unloading, multimodal services and some value-added services were provided	Cargo and information flow intensified packages of comprehensive services. Highly value-added services were also provided	Priority is given to staff training, improving port service qualities (making operations and management more people-oriented)
Decisive factors	Labor and capital	Capital	Skills and techniques	Information technologies

Source: UNCTAD (1992) and Beresford et al. (2004).

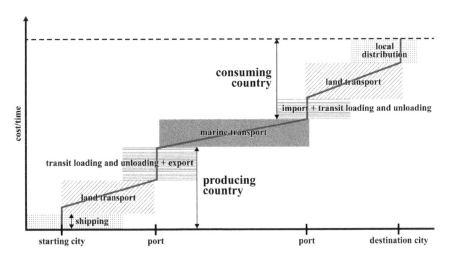

Figure 2.1   Port in Multinational Products Supply Chain

looking into micro spatial setting of port facilities. In this sense, the basic and overall function of port will never change: ports are the maritime interfaces with loading and unloading functions, bridging various transport modes for goods and people. No matter if it is a transfer between ships, or a transfer between land vehicles and ships, essentially the transition interfaces are always ports.

This interface between unmatched transport links frequently results in the different transport carriers being unable to match each other perfectly in time and space and results in unavoidable geographical interruptions in the transport chain. These pauses and transitions incur not only extra costs for loading and unloading, consolidation and deconsolidation, but also extend the lead times and the costs of temporary storage (see Figure 2.1). Therefore, areas near ports have become the locations to set up warehouses and naturally became locations chosen by other logistics value-added enterprises for carrying out their production activities. In this way, new extra geographical pauses are avoided.

The line of business and physical extent of ports are related to ways of handling relevant resources and government and corporate activities. For casual observers ports are seen as bays where ships pause, load and unload, and come and go. For them the exact geographical boundaries of ports, such as Shanghai, are unclear, nor are they sure whether "Victoria Harbor" in Hong Kong is a port or not. However, for the national government, "coastline" is a kind of public resource and port means a certain boundary for occupying and using this kind of resource. At the same time, as a transportation link, ports are closely connected to relevant businesses and spaces in terms of functions. Therefore, ports must have a clear definition to facilitate management. We need to note that the definition of a port involves the following three basic operating units: berth, terminal, and pier, although the latter is often loosely defined to refer to either berth or terminal. It is

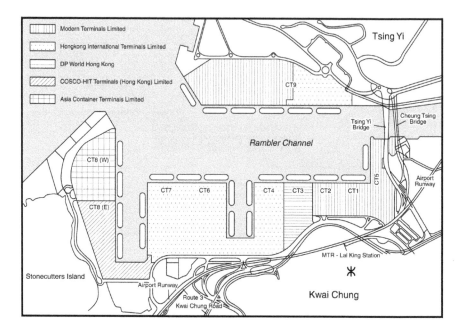

Figure 2.2   Schematic Drawing of the Kwai Chung Container Terminals

easy to understand berth. It refers to a constructed facility with a certain length designed to allow vessels tie up alongside the shore. Piers usually only handle a certain kind of goods (such as coal, oil or containers) or a group of passenger berths and the relevant onshore facilities. Meanwhile, in Chinese, the word pier, like port, has a broad meaning. It often refers to all of the land used for port business. A unit of terminals exists in port management vocabulary. In Chinese, its accurate translation should be "operation area (作业区)", but sometimes it is also translated as "pier (码头)" or "port area (港区)". Pier operators often use the plural form of terminal in their companies' names to indicate that they operate in multiple terminals, and each terminal may contain several berths.

To illustrate the above situation, we can take Hong Kong as an example. In Hong Kong, the business of container handling is mainly concentrated in the Kwai Chung-Tsing Yi area, usually called the Kwai Chung Container Terminals. As Figure 2.2 shows, there are five terminal operators (see symbols in the upper-left corner), nine terminals (CT1-CT9), and 24 large container berths (marked by the ship shape on the map). In addition, Hong Kong has other terminals outside the Kwai Chung Container Terminals, including the River Trade Terminal and water anchorage, the so-called "mid-stream terminal" (see Figure 2.3).

There are two reasons for explaining the concepts of ports, piers and terminals. First, from the actual layout of berths, piers and terminals, it can be seen that the space composition of ports is highly fragmented. The space components can be a

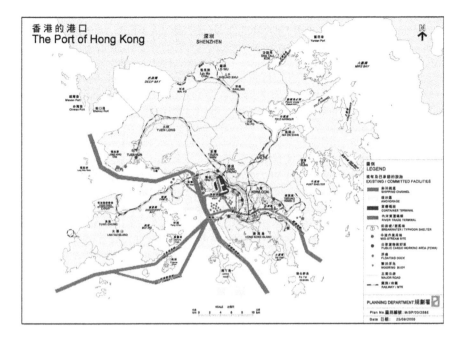

Figure 2.3   **The Port of Hong Kong (including the Kwai Chung Container Terminals)**

whole unit (such as the alongshore area of Pujiang in Port of Shanghai), and they can be fragmented. Therefore, "ports" can be composed of several fragmented parts. Second, although services and facilities related to customs, navigation and passages can be operated for the port as a whole from a management perspective, the basic operations of berths are conducted by the terminal unit. Sometimes, a terminal independent in space can also be called a port (such as Yantian Port – some consigners in the United States and Canada have not even heard about the Port of Shenzhen). Therefore, "port" is not the sole statistical or administrative unit reflecting its core functions from a physical or management perspective. This directly involves **the following second set of basic concepts: the definition of a port city.**

A port city pertains to a city with a port. Two points should be addressed before examining port cities in China. First, English literature uses two different concepts, namely port city and city port. Port city is similar to the Chinese sense of the word, that is, a city that also functions as a port. A city port refers to a city which has a port for its primary growth mechanism, such as Rotterdam in the Netherlands and Le Havre in France.

Second, cities in China must be regarded as an administrative division in consideration of the extractive political institutions of the state, which remained as the dominant driving force behind the economic and social changes in the People's Republic despite the economic reforms in 1978. Three types of city administration are employed in China: a municipality, which is a provincial-level division (e.g., Shanghai); a prefecture-level city, which is governed by provinces or autonomous regions; and a county-level city, which is a sub-unit of an administrative division at the prefecture level. Hence, the relationship of a port to a city does not simply entail the physical occupation of transport infrastructures and the location of their operations along the waterfront of a coastal city. This topic is further discussed in Chapter 6. Every port authority in China is bestowed with administrative power by the Ministry of Transport (previously the Ministry of Communications), which is one level lower than the power conferred to the city where the port is located. Thus, the administrative power of a port in a city in Mainland China is always associated with the power of its city. For this reason, the methodologies used in analyzing Chinese port cities in this book are distinguished from those used by Hall and Jacobs (2012), by taking into account of spatial jurisdiction more seriously. This approach is important because terminals and port areas are statistically located in a particular city jurisdiction, which are normally counted as one port. The ports and their sources of relevant infrastructure funding or government subsidies can be related directly to the development of the city. To some degree, this situation is similar to that in Europe, where some ports are not far from one another. However, these ports are supported by different countries given that they belong to different countries. Thus, port city clusters are developed.

After the implementation of the policy of "developing counties through cities" in some cities in China, a port established in a county-level city became a part of the main city. For example, Port of Jiangyin, which is located in Fuqing in Fujian province, was converted into a port in Fuzhou after Fuqing became part of Fuzhou. This case shows that city boundaries and any changes thereof influence the changes in port location and development opportunities, particularly when territorial rescaling occurs through the merging of ports with a city of a lower administration level.

The third set of basic concepts is related to port resources. With economic globalization, ports are increasingly regarded as a kind of resource. Investors have gradually begun to view ports as investment targets because they can bring stable economic returns. What kind of resource are they, and how can they bring satisfactory returns? First of all, ports have a coastline and ship channels. The combination of deep water coastline and deep water channel is a special kind of "**trade interface resource**" with a geological monopoly. Although it needs other conditions to transform opportunities into reality, the ownership of this kind of resource means more opportunities to become a part of the global trade system. However nowadays, technology is highly developed; it is possible to build artificial deep water channels and deep water ports. The key is cost. Therefore, different port cities have different development opportunities due to different natural resources and financial conditions.

1. Some port cities such as Manchester in Britain and Guangzhou in China once developed foreign trade through ports, but as ships became bigger

and the requirement for water depth became higher, this kind of **trade interface resource** gradually disappeared in the areas administered by the city. As a result, they either connect themselves with other international hub ports in the form of branches to conduct international trade, they dig deep water channels to change existing ports (for example, Tianjin), or they build deep water ports across administrative regions in nearby areas (for example, Shanghai has built the Port of Yangshan in Zhoushan) to solve problems. In special cases, after the expansion of administrative borders, the trade interface resource comes back.

2. In the second kind of city (such as Qingdao and Dalian), although the existing port is not deep enough, there are other locations with deep water in the administrative boundaries. Therefore, the locations with deep water are more developed than existing ports (such as the development of Qingdao Qianwan and Dayaowan of the Dalian Port).

3. The third kind of city is that with the coastal conditions within its jurisdiction that can be used as **trade interface for maritime shipping,** such as Hong Kong, Shenzhen and Singapore. Combined with other development conditions, the resources have become ideal international ports.

4. The fourth kind of city is the coastal city (cities along rivers) with excellent conditions for deep water ports but lacking other conditions (such as Fujian Meizhouwan and Guangdong Zhanjiang). As China plays an increasingly greater role on the international stage, the development requirement of these "interface resources" has been put on the agenda.

5. Finally, although many coastal cities or cities along rivers have had their own ports for many years, conditions like water depth cannot meet the requirements of "**trade interface resources**". Furthermore, this problem cannot be addressed with existing financial and technical conditions (such as in Manchester, and Quanzhouwan in Fujian). The fact that ships have become larger over the past three decades (see Figure 2.4) is the main reason ports are nearer deep water coastlines.

Highly related to the trade interface resources is the concept of **transport gateway**. In the existing geography literature, gateway is generally defined as a transportation hub, such as seaports and airports (Pain 2007), a pivotal point (Mason 2007; Tretheway, Andriulaitis et al. 2007; Rodrigue and Notteboom 2010), or a node that provides the entrance to and the exit from its hinterland (Berechman 2007; Mason 2007; Pain 2007; Rodrigue and Notteboom 2010). A gateway commonly implies a shift from one mode to the other (Rodrigue and Notteboom 2010). Its characteristics have been explored by many scholars (Bird 1971; 1983; Burghardt 1971; Van Klink, Arjen et al. 1998; Button 2007; Gillen 2007; Hall 2007; Mason 2007; Rodrigue 2007; Tongzon 2007; Tretheway, Andriulaitis et al. 2007; Williams.G.Morrison 2007). To synthesize these characteristics, a gateway can be generalized as a national or regional pivotal node with a protocol/translating function, and is located in a strategic site with supporting infrastructure and

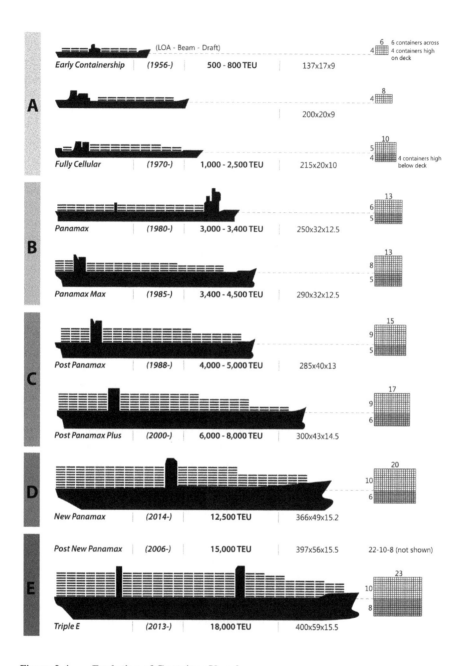

Figure 2.4   Evolution of Container Vessels

Source: Adapted from J.P. Rodrigue et al. (2012) Geography of Transport Systems (online version)

services to interface systems of trade and/or transport with different regulations for the region it belongs to. Thus, many port cities in China are the gateways to the rest of the world, and as a result, port cities and their growth are regarded by the state as a national development strategy.

In addition to "**trade interface resources**", port cities in developed countries, such as Sydney in Australia, Liverpool in Britain, Marseilles in France, Venice in Italy, Philadelphia and Baltimore in the US and Hong Kong and Singapore in Asia, are experiencing another common trend – the transformation of the original port area into a waterfront commercial/residential/tourist district, a process generally called "waterfront redevelopment". Behind this trend are two core driving factors. The first is the above-mentioned shift to deep water terminal operations, which has caused the slowing of development in the original port area; the other is the requirement for **urban waterfront resources** by the city itself. Recently, Shanghai invested tens of billions RMB to transform both sides of Huangpu River, which shows that the trend of waterfront redevelopment is becoming popular in China. Complementary with "**trade interface resources**", **urban waterfront resources** can increase local attraction with local features. **Trade interface resources** advocate that port functions should be expanded through direct exchanges with the international marine trade. In contrast, **urban waterfront resources** stress the "experience economy" and the value of land. In other words, they stress the commercial values that the coastline or other waterfront areas signify for residence, tourism and leisure activities. For famous ports, urban development often revolves around the old port area. Therefore, the old port area has not only become a part of the downtown area, but also is often surrounded by historic sites, which further improves the non-port functions of the old port area. To some extent, the more famous a city was in history as a trading port, the more valuable it is for conducting high value-added activities in its old port area. However, not all ports' coastlines can be transformed to the above-mentioned waterfront resources. After all, their "selling points" are different from the time when the ports were developed.

2.2 Historical Background of Port City Development: Evolution of Trade Patterns and Marine Technologies

Port cities are developed for many reasons. These reasons can be classified under four major categories, with each type represented by its resulting cities. The first type includes ports that are formed by the demands of local trade and then gradually (but not necessarily) expanded to involve regional or even international trade, such as the port cities along the Yangtze River and those in Britain, Holland, Japan, and South Korea. The second type includes ports formed from the exploration, emigration, and development in the "new world". These ports represent a colonial relationship such as the ports in New York in the United States, Montreal in Canada, and Sydney in Australia, which were built with the aim of developing the new regions. The third type includes ports that are formed from colonial plundering and the development of other land resources, such as the ports in Cape Town in South Africa and Hong Kong in China. The fourth type includes ports that are initially built as technically necessary stopovers for refueling and maintenance, or as short-cut passages for long-haul marine transportation, such as the ports in Panama in Central America and Singapore near the Malacca Strait. These intermediate ports are constructed in a strategic location. Some of these ports may eventually become a place with centrality if the economy of the city surpasses the role of an intermediate hub for cargo shipping.

Regardless of how a port city is formed, once the port becomes involved in the international economy, its development becomes affected by international trade patterns, marine technical advancement, and the operational organization. Changes in the international trade pattern, which refers to the contact between countries and the relationships behind trade relations, involve unbalanced global economic development and technical advancement. Before the twentieth century and in the early twentieth century, non-market and even non-economic means such as feudal empires, colonial expansion, and resource plundering usually had more impact than normal market trading. Thus, the formation and evolution of port cities in typical colonial countries were different from those in the European sovereign states. Furthermore, the dynamic imbalance in resource distribution and product production levels resulted in continual spatial changes in complementary and demand and supply relationships.

Given the above-mentioned dual relationships, the importation of raw materials and exportation of industrial products were limited to Commonwealth nations and colonial countries. The exportation of industrial products from Manchester, located at the center of the Industrial Revolution, and the importation of raw materials from colonial countries in Africa transformed Liverpool, a city in Western Britain, into a leading port in Europe at that time. Industrial European countries such as Holland and France, which regarded overseas resources, including their territory, as their future, built and developed port cities in their own land as well as in their colonies across the world, such as in African and Asian countries. Ports established in other countries such as bridgeheads include Qingdao and other Chinese cities, which were established for exploiting inland resources and/or opening new markets.

An established port city experiences many triumphs and difficulties. Its triumphs are not necessarily related to its functions as a port (e.g. New York), although its difficulties are usually related to the loss of its historical glory, which may be due to the rapid development of modern marine technologies. For example, due to the appearance of large container ships, the canal between Liverpool and Manchester became useless for any economically meaningful size of cargo shipping. The role of port of Liverpool has changed drastically due largely to two major reasons. First, internally with the UK, as argued by Hoare (1986), both demand patterns for maritime shipping and the land connections with various ports in the UK changed substantially. Second, externally, the global economy and its trade pattern also shifted significantly after World War II. Indeed, the globalization trend in the last three decades has broaden the trade horizon to more newly industrialized and industrializing countries, and as a result, marginalized the role of British ports in general in the global maritime network today.

Low cost shipping indicates the shortening of the absolute cost distance between ports and cities. Moreover, low cost shipping also refers to the change in the relative cost distance between port cities and other cities and regions in the sense of "time-space convergence". For example, shipping a 29-inch color TV only costs USD 15 (RMB 125) in freight charges from a factory in Dongguan City in Guangdong Province to an appliance store in New York City in the United States. This cost is much cheaper than the cost of shipping the TV to an inland city somewhere in Western China. As such, firms inevitably relocate their production lines to countries or regions with lower labor costs and land rent, provided that the requirements for the production time are not so high, few man-made trade barriers exist, and enough demand exists for the product in connected foreland markets. Such an arrangement provides incentives for coastal cities in developing countries to develop their ports and reduce the distance from global market places, enabling these cities to fight for opportunities and their share in the global economy. Therefore, port cargo handling capacities in developed countries have rapidly expanded with the globalization of production. The role and position of the logistics industry in many port cities have also changed accordingly.

2.3 Background of the Development of Port Cities: Global Economic Integration

Since the 1970s, particularly in the recent decade, the concept of the market economy has been increasingly recognized and adopted by more and more countries: international trade barriers have gradually broken down, the power of transnational enterprises has strengthened, and the trend in global economic integration has become increasingly clear and irreversible. Global economic integration pertains to four international trends: (1) the manufacture and sale of commodities are arranged and organized globally, resulting in the rise of the total transnational and transregional physical goods flow volume; (2) transnational transmission and mutual influence of various information; (3) transnational flow and transfer of capitals; and (4) more people flow due to the social, economic, and cultural activities conducted globally. Transnational flows are directly related to the development of sea, land, and air transport technologies and the logistics industry, as one important condition of economic and trade globalization is the internationalization and integration of transport-logistics facilities and services.

Nevertheless, different modes of transportation have resulted in different improvements in quality, quantity, cost, comfort, safety, and management. Even for a particular transportation mode, operational characteristics, management approaches, and technological levels vary for different locations. Therefore, the actual conditions of accessibility in different places in the world vary considerably and so does the rate of progress. As a result, the degree of participation and the competitiveness of transport systems in the world economy differ for various countries and cities.

The reason why some cities and countries are able to participate more or gain more in a globalized system is not only because of their open economy and society but also because of their high accessibility from other places in the world through modern transport infrastructures. For example, intercity personnel mobility is highly associated with the accessibility provided and influenced by the service level of air transport and individual airports. With regard to the flow of goods, ports with a worldwide maritime shipping network are among the most important nodal transport facilities and infrastructure for global trade. Indeed, the main reason why the major ports of the global economy are concentrated in urbanized areas by the sea or along rivers is the low-cost, natural transportation achieved via water-borne shipping. Consider the recent rise of Brazil, Russia, India, and China (BRIC). Three of the BRIC countries experienced enormous growth in their port business, reflecting the deep relationship between port development and globalization. This phenomenon was evident in the most dramatic change undergone by China: the total container port throughput for the whole country increased by 50 percent annually within three decades (from 1978 to 2007), and Shanghai ranked as the number one port in the world in terms of total tonnage. More importantly, from the perspective of a port city, the coastal regions (i.e. areas that are 150 km from the coastal port cities) accounted for 92 percent of China's total foreign trade value. In other words, while the seaport system helped this country to maintain its rising economic growth at about 9 percent annually for three decades, such a system also led to a very high concentration of international trade geographically biased towards the coastal cities. Thus, China has been divided into two: a containerized China (i.e. involvement of coastal port cities and their surrounding areas in export-orient manufacturing activities) and a China yet to be containerized (i.e. only regions involved in international trade are containerized). Within containerized China, the Yangtze River Delta and Pearl River Delta are two of the largest geographical concentrations of manufacturing factories and logistics activities in the world. These regions include two major ports of Shanghai and Ningbo of the Yangtze River Delta and the three major ports of Hong Kong, Shenzhen, and Guangzhou of the Pearl River Delta.

Therefore, the fact that cities compete with one another through the continuous improvement of their port connectivity and gateway functions to the world economy is not surprising. However, comprehensive analyses on port-city relations remain limited, although organizations such as the International Association of Cities and Ports have exerted consistent efforts to examine the individual aspects of port cities (e.g. AIVP 2008). Specifically, many relevant key issues are not supported by mature theories; even experts are still working on such theories. Therefore, this book is an attempt to examine the issues through a novel and structural perspective. The interactive relationships between the ports and cities in China are considered actual examples to address some issues observed. These issues are as follows:

- What are the dynamics of the functional, economic, and spatial relationship between a port and its host city?
- With respect to global economic integration, has the nature of ports changed in fostering the cities where these ports are located?
- What is the relationship between multimodal transport logistics and the development of ports? How are port cities involved in the formation and governance of a multimodal transport gateway centered at the port?
- How do international port operators influence a port and its host city?
- What is the role of the individual city government in case of a division of labor among ports in the same region? How does the regional or national government act in such a situation?

2.4 Analytical Framework of Port-City Relations

The aim of building the analysis framework of port-city relations is to compare between cities that have greatly benefited from their port development to determine whether such benefits are natural or incidental, as well as to identify the driving factors behind these benefits. Ports and cities have different characteristics. On one hand, ports are the origin of the development of some waterfront cities and the source of their continuous development. On the other hand, cities have their own development trajectory and driving engines. Even port cities have different degrees of dependence on ports due to various internal and external reasons. Nevertheless, port cities must have something in common. A good analytical framework can reveal these common aspects and analyze port cities from multiple perspectives. Such framework will enable us to distinguish the features and differences of port cities and to identify common interactive processes and systems for port-city relationships.

With the unimaginable complexity of the real world, port-city relations have many aspects, such as functional, spatial, financial, interpersonal, and organizational relations. Many cross-points exist in these relations and constitute the port-city interaction, and these points cannot be easily isolated for any single relation. Nevertheless, we still need a framework that can identify such connections to reveal the main points in the interactive relationship between ports and cities. These relationships can be summed up by four aspects or axes from the perspective of port cities.

The first aspect is the economic and functional relationship. As an interface connecting different transportation modes, a port first influences the city where the port is economically positioned and located. However, the economic contribution that a port can make to a city may vary, as this economic distribution is related to the role played by the port in the transportation system and the economic structure of the city in which it is located. Furthermore, the relationship between the port and the city is always changing. The economic relationship involves the needs of the city for the port, the influence of the port on industry development, the investment made in the port by the city, employment opportunities generated by the port, and contributions made by the port to the overall economy of the city.

The second aspect is the interactive geographical relationship. Ports and cities either gradually develop or become less prosperous. Regardless of the situation, the spatial pattern of a city and the relationship between the city's different parts influence the port operations in the city. Moreover, the location of the port can either limit the possible future changes in the city or result in different directions of development. The selection of the location of a new port is related to the urban environment and pattern. Moreover, the interactive process of the expansion of ports and cities and the waterfront redevelopment are also important aspects of the port-city spatial relationship.

The third aspect is the external network a port brings to its city. A port is a node for transportation. Therefore, the marine network of a port can be expanded once the port attracts a large number of shipping lines. With the advanced shipping technologies and the globalized trade and production, this kind of network can bring unimaginable development opportunities to the port's home city. The connections formed by various modes of transportation to other places also transform the port city into a place of intermodal transport hub, which can potentially attract more cargo transshipment as well as value-added trading and logistics activities to the city. In turn, enhanced comprehensive logistics services and a higher level of network connectivity may reinforce the port's strategic location for more shipping and port business.

The fourth aspect is port-city governance or the coordination and managerial relationship between the port and the city. This relationship involves matters such as land and port ownership, the holding structure of managerial power, planning participation, and mode of administration of the city government in relation to various levels of the port business. With the participation of international terminal operators, the changes in the national port management system, and the changing port governance, this kind of relationship has become increasingly complex and diverse.

These four aspects are interrelated and influence one another, although such information is not enough. As no three-dimensional chart can fully illustrate the complexity of the relationship, we use Figure 2.5 to demonstrate the characteristics and interactions of port-city relationships through the four aspects. Economic and functional relationships are placed at the center to indicate that port functions are at the core of the relationship between the port and the city where it is located, reflecting the demand–supply relationship between the port and the city. If a port is only a transit point for bulk products (e.g. a large coal-exporting port), and the city where the port is located does not need or produce coal, almost all the clients of the port may be located outside the city. In this case, the port-city economic relationship refers to the limited economic contributions made by the port. Conversely, if the port is a large container port, a cluster of enterprises that supplies containerized cargos may be found within the city. In such case, the economic interactive relationship between the port and the city, as well as the

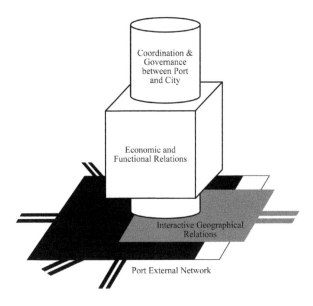

Figure 2.5   Four Main Relations Between Port and City

contribution of the port to the city, can be significant. The governance relation is placed at the center of the economic relationship, indicating its function as an axis penetrating every relation to highlight the city's participation in the development of the port and in coordinating the port-city relation. This function can directly change the pace and direction of port development. The port-city governance also influences the geographical changes in the port city through policies and planning. The interactive geographical relation is indicated as a layer beneath economic and functional relations to show that this relationship reflects the physical realization of the port-city economic relationship. At the same time, the geographical relation is the ultimate reason for us to take an interest in port development from the perspective of urban planning and development. The external network relation under and around the three other relations is placed at bottom to illustrate the exterior nature of the port-city relationship: it is not subject to the influence of the port-city governance but rather to the city's external connections obtained through port development. The intensity and characteristics of this network connection reflect the foreland market, hinterland, and development potential of the port city as a gateway for the whole country or region, as well as its role in global economic integration. For this reason, the external network relationship has been included in the port-city study.

Apart from the four relations, the development of a coastal port city is usually influenced by higher-level systems. In China, the Ministry of Communications (currently the Ministry of Transport) used to plan, build, manage, and operate ports directly, as ordered by the central government. Since the 1980s, privileged policies

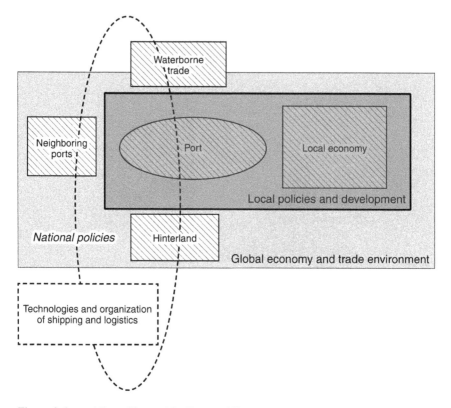

Figure 2.6   A Port City and its External Factors

resulted in 14 coastal port-cities being more economically advanced, as they have enjoyed the macro-market-environment earlier than other places in the country. From another perspective, these 14 cities were chosen mainly because they could be connected directly through port and shipping networks to the outside world, both the major markets of consumer goods and the major natural resources to produce them. In this regard, the four abovementioned relations can be considered the internal port-city relations. As such, a larger scale relation still remains to be seen in the external environment where port cities develop. Figure 2.6 illustrates how a port city fits in its external factors.

Figure 2.6 presents the key external factors or exogenous variables of the port-city system. These factors can influence specific port cities with regard to the economy, governance, space, and network interacting with any individual port and its city. More importantly, such a macro-external environment for port city development may be generally similar in terms of the factors involved, such as global economy and trade, technologies used in shipping and ports, organizational changes accompanying these technologies, and the competition or cooperation from neighboring ports and cities. However, this external environment may vary from city to city, with varying degrees of involvement. One of the main tasks of this manuscript is to provide a detailed account of how such a complicated macro setting interacts with the growth of and change in port cities in China, a country that has recently experienced a drastic changes from being a centrally planned economy to a more market-oriented, internationally open economy in the past three decades while maintaining the strong hand of the state, which intervenes in almost every aspect of the country.

The natural environments of the port and the city are not included in this framework in order to reflect the reality in contemporary China for the past three decades. Instead, the development driven by economic growth has been the focus. Such development has resulted in the oversight of issues in environmental protection and sustainability. This total disregard for sustainable issues will be discussed in Chapters 8 and 9.

2.5 Summary

This chapter sets up a foundation for the analysis of China's port-city interplays. It provides some basic notions first for readers to understand some special background regarding, for example, how port and city are associated in space jurisdictionally. It describes then the globalization in the past decades and how it involves with the evolution of port cities in China in the last three decades. Studies on port cities conducted in other places can still be used as reference, but they cannot provide a reasonable and complete picture of the reality in China. Last, this chapter proposes an analytical framework of a port-city relationship composed of four aspects, namely, economic and functional relationship, interactive geographical relationship, port-city governance, and external networks. It is argued that to decode the port-city interplays in China so far, this framework is most suitable in revealing the growth-driven evolution of port-city relations under strong intervention of the states. The following four chapters explain each of these four sets of port-city relations in detail.

Chapter 3
Port-City Economic
and Functional Relationship

3.1 Economic Contributions of the Port to its Host City

Similar to those in many other countries, coastal cities in China have developed faster than inland cities over the past three decades. Coastal cities with larger geographical scales are more economically developed than those with smaller geographical scales (Shi and Hui, 1996). However, measuring the economic contributions of a port to its host city is difficult. If the typical economic influence measurement is applied, we are required not only to separate other factors, but also to calculate the direct and the indirect economic contributions of the port to the city.

Direct economic contributions refer to the initial or to the first-round effect of the port service industry to the regional and to the national economy. The port service industry is a business that directly performs cargo shipping around the waterfront, including loading and unloading, piloting, barging, storage, and cargo consolidation and deconsolidation. As its core, the port service industry operates under the jurisdiction of the Port Management Bureau, and its added values and job opportunities comprise the most direct part of the port's contributions to the city. Figures for these contributions are easy to obtain. The related industries of a port, including attracted and derived industries, consist of enterprises that can gain commercial benefits from the entry and the exit of goods in a specific port. These industries have significant importance to the convenient transport of their products to the port. As road transport costs much more than sea transport, these enterprises usually consider geographical proximity to the port. The added values generated by these enterprises are regarded as business secrets and are therefore difficult to calculate accurately. As a result, some scholars calculate the annual added value of related industries using the correlation coefficient between industries and ports. In 1998, the Shanghai International Shipping Institute and the Shanghai Statistical Bureau developed a model that used the port-industry correlation coefficient to analyze the direct economic contributions of ports. This model collected data from 1693 enterprises from different industries that include transportation, commercial, construction, financial and insurance, and social service industries. The added value generated by the shipping industry and its related industries in Shanghai was 19.312 billion yuan, accounting to 5.7 percent of Shanghai's GDP (336.021 billion yuan).

Based on the above calculation, Song (2000) further measured and calculated the indirect economic contributions of the Shanghai Port using the input-output matrix. As a typical method to estimate the multiplier effect, Song defines the indirect economic contributions of ports as second-round economic effects, brought about by the spread effect of the first direct economic impact of the port service industry and its related industries on the regional economy. The total added value is thus calculated by using the multiplier effect in the input-output matrix, that is, the sum of the second to the last rounds of economic effects (reduced to 0) of the port and its related industries on the regional economy. Song's calculation revealed that the added value of the indirect contributions of the Shanghai Port amounted to 11.812 billion yuan. Along with direct economic contributions, the total economic contribution of the Shanghai Port to the regional economy was 31.123 billion yuan, accounting to approximately 9.3 percent of Shanghai's GDP that year and generated about 700,000 job opportunities. The result of the calculation indicates that the direct economic contribution of the Shanghai Port accounted to less than two thirds of the total economic contribution, whereas the indirect economic contribution of the Shanghai Port was more than one third of the total.

If the above calculation on the economic contribution of the Shanghai Port is not exaggerated and if the influence of ports on cities has remained since 1997, the economic contribution of ports to China's coastal cities will be 6 percent of the total national economy as coastal cities generate more than 70 percent of China's GDP. However, the situation is more complicated. Port-related industries vary among different cities. Consequently, the contributions of ports to cities differ. This variation requires us to examine the functional relationship between ports and cities in which the port-city economic relationship is premised.

3.2   Functional Relationship Between the Port and its Host City

A seaport is a land-sea interface for the flow of goods and passengers. The flow of goods and of people in a port can be divided into four categories:

1. Intermodal through-transportation that land-connects the hinterland internally and water-connects coastal and overseas cities externally for industries such as
    a. Cargo-handling businesses
    b. Storage
2. Sea-borne transport for coastal processing industries
    a. Port services for processing and for export industries (e.g., shoe manufacturing, apparel processing, and electric appliance assembly)
    b. Port services for local needs, such as power stations, and the food processing industry
3. Vessel-to-vessel transshipment of cargos
4. Traveler services (e.g., cruises and ferries)

The first three of the above categories are related to cargo transport. Figure 3.1 demonstrates the functions of different ports and the relationships among them, their host cities, and their corresponding hinterlands.

Figure 3.1 shows that types I (through-port) and III (transshipment port) have a relatively simple relationship with the host city. The economic contributions of these ports are confined to their most basic functions. For example, Qinhuangdao is a typical type I port where large amounts of coal and crude oil have been transported between land and sea for many decades. However, the city neither produces nor processes coal and oil. Once relevant oil or coal-processing enterprises settle in the city, the port becomes a combination of types I and IIb. As a result, the port's indirect contributions to the city increase. Similar situations may also exist in ports that conduct transshipments, as value-added activities accumulate due to intensive connectivity to other places that is strengthened by transshipment activities.

Ports that belong to either type II-a or type II-b have a close relationship with their host cities. The former represents a typical situation in which processing and manufacturing industries are concentrated in a coastal city. The latter demonstrates a situation where basic industries, such as oil refinery and power generation, require the bulk import of raw materials from abroad, resulting in the seaport as the most suitable location for the provision of basic needs of both local and inland areas. Products can be transferred to inland areas through various modes of transport, including pipelines and electric networks.

Figure 3.2 and Figure 3.3 respectively illustrate the statistical relationship between GDP and container throughput and the statistical relationship between the two in eight mainland Chinese cities from 1986 to 2005 against the same data from Hong Kong from 1976 to 1993. The curves show that cities with different economic structures are influenced differently by ports. First, the situation in Shanghai is very similar to that of Guangzhou in terms of the relationship between

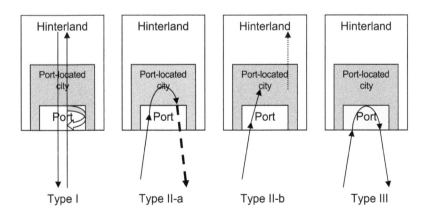

Figure 3.1   Goods Shipping Functions Categorized by Relationships Between the Port and its Host City and the Hinterland

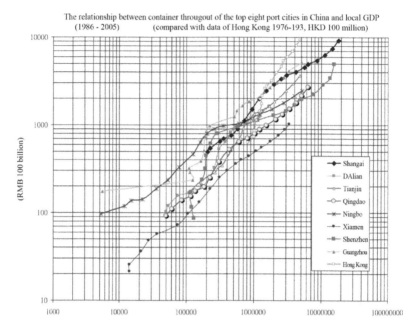

The relationship between container througout of the top eight port cities in China and local GDP
(1986 - 2005)   (compared with data of Hong Kong 1976-193, HKD 100 million)

Figure 3.2   Relationship Between Container Throughput of Top Eight Port Cities in China and the Local GDP

Source: China Infobank, Census and Statistics Department, Government of Hong Kong Special Administrative Region, Institute of Comprehensive Transportation, National Development and Reform Commission

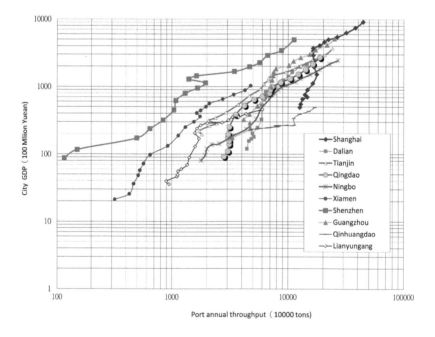

Figure 3.3   The Relationship Between Port Throughput of the Top Ten Port Cities in China and the Local GDP (1986–2005)

container throughput and GDP. The increasing rate of containers started declining when GDP reached 200 billion yuan per year. Guangzhou accomplished only two thirds of the achievement of Shanghai (in the same period of time), which took 15 years to attain their economic scale and level of container throughput. Second, Shenzhen and Qingdao seem to be in another group with similar experiences on the ports' influence. The increased interrelationship between economic growth and container throughput in Qingdao is very similar to that in Shenzhen. Meanwhile, Qingdao reached GDP of 200 billion yuan per year after two to three years. When the GDP of Shanghai and Guangzhou reached 200 billion yuan per year, both handled 1.25 million TEU annually. By contrast, Shenzhen and Qingdao handled an annual number of 5.2 million TEU when their GDP reached the same level.

If we have good knowledge of the comprehensive functions of Shanghai and of Guangzhou as core cities of China's two most economically developed deltas and of Shenzhen and of Qingdao as the most important processing export bases and key container ports in their host regions, the implication of the different influences of ports on these two groups of cities will be easy to understand. The economic contributions of a port to a city are largely determined by the city's economic structure. This observation helps explain the categorization of port cities in Figure 3.1. The container throughput of port cities such as Qingdao, Shenzhen, and Xiamen generates relatively low GDP. The ports of these cities can still expand their economic contributions to their host cities because their export- and processing-based economies are highly dependent on large-scale container ports compared with Shanghai and Guangzhou that are metropolises with diversified economic structures. Moreover, if the trade through container shipping is seriously unbalanced, container throughput merely represents a large number of empty boxes that slightly contribute to the city's economy.

Different port cities are difficult to compare by examining the curves about the ports' influences on their host cities, with GDP and port throughput of all cargos as variables. Comparison is difficult because the average value of all cargos in a port is much different from that of all containers in another port. Some ports handle cargos for local mass consumption, whereas other ports simply serve as a hub for transshipment or as a gateway for raw materials, such as coal and oil. Therefore, port cities should be divided into different categories to clearly distinguish their influences on their host cities.

3.3 Dynamic Economic Interaction Between the Port and the City

Apart from using throughput and GDP as variables to analyze the economic relationship between a port and its host city, the question remains on the possibility of further examining the interactive economic relationship between a port and its host city, such as on the relationship between a port's handling capacity and demand and on the existence of an "ideal" port-city economic relationship. Chen (2006) investigated the relationship between the port and the city in Dalian as two systems using econometrics to demonstrate the interactive and the coordinating process of the relationship between the port and the city economy. This research establishes two sets of indicators representing their corresponding economies, and then implements the principal component analysis to regroup these indicators. The two most significantly contributing and independent sets of indicators are selected from each system to represent the respective economy. The scale of the port (share of contribution: 64 percent) and the foreign trade throughput (share of contribution: 18 percent) are two primary indicators of port economy, while economic strength (share of contribution: 86 percent) and economic openness (share of contribution: 7 percent) are the two principal indicators of city economy. Finally, a "coordination index" between port and city economy from 1990 to 2003 is generated through fuzzy logic models of mathematics (Chen 2006). The outcome of this research reveals that the port progressed faster than the city in the early 1990s and then their growths became more coordinated in the subsequent ten years. This study is a meaningful attempt because it performs an insightful examination of the increasingly coordinated dynamic relationship between the port and the city. Unfortunately, after the transposition of the principal indicators, the "coordination index" reflected changes on the extent of the port's basic operation (as reflected by port throughput) in serving the city rather than on the level of the "port economy", that is, the value added directly and indirectly by the port. This value should be the real economic contribution of a port to its city. Moreover, given the unavailability of data, no research has demonstrated the interactive relationship between port economy and related industries in host cities in China. On this regard, Chinese port cities are less quantified in comparison with similar studies done, for example, for some European cities such as Hamburg in Germany (Merk, and Hesse 2012).

3.4   Evolving Port-City Functional Relationship

If a port develops under ideal conditions, not only does the amount of its specific or primary cargos (e.g. bulk cargo or containers) increase, but also its scope of service expands. Large-scale ports, such as Shanghai and Guangzhou ports, can have all the aforementioned functions and demonstrate complicated relationships with their local areas and corresponding hinterlands. These ports also have a close relationship with the host cities of other coastal ports. Consequently, the contributions of the port to the economy become difficult to measure. With time, the functions of a port evolve parallel to the development process and to the functional transformation of its host city (Figure 3.4). However, the world is ever-changing. Whether a port's function gradually evolves and whether its host city undergoes every development stage remain uncertain.

Such uncertainty is caused by many factors. For example, the large-scale spatial transformation of industries directly related to a port may alter its development from the originally anticipated trajectory and functional upgrade. Liverpool, for

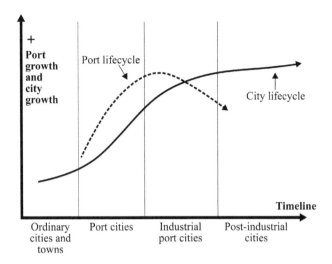

Figure 3.4　　**Functional Evolution of a Port and its Host City**

instance, once became the most important port between Africa, Britain, and the European continent after the first industrial revolution. However, this port lost its glory with the changing global trade relations. The upgrade of port functions and scale alone cannot maintain the previous contributions of a port to its host city. Thus, Liverpool as a port became weak for several decades. Similarly, Quanzhou Port lost its status as the largest port in China as a result of Fuzhou becoming the capital after a series of anti-government riots in Quanzhou some 300 years ago. Moreover, some port cities, such as New York and London, became increasingly independent from their ports as they developed. This case is different from those of Liverpool and of Quanzhou that were "once glorious" but "fell halfway". London and New York City upgraded themselves from a port-based trade hub of tangible goods to world-leading hubs for intangible goods or as financial centers. Hong Kong seemingly followed in the footsteps of these two cities. The functional transformation and upgrade of a port may not be in tune with that of its host city and may also have different life cycles (Figure 3.4).

The two life cycles shown in Figure 3.4 do not coincide, thus revealing that the reliance of a port city on its port may change as the city develops at different stages. Using econometric models, Japanese regional economist Fujita verified that the influences of related service industries, such as the financial industry, as a result of the growth of a port remain rather than decrease despite the spread of port activities to other places and the reduction of the total volume of port activity (Fujita 2005). This remaining influence is the key to the structural transformation of the city economy. We explain the different life cycles mentioned above by considering Hong Kong as an example.

3.5   The Case of Hong Kong: From a City of Processing Trade to a Global Financial and Service Center

Hong Kong was once a fishing port, a commercial port, and a treaty port in which Western countries traded with China. Hong Kong became a British colony in 1841. When the People's Republic was founded in 1949, Hong Kong remained as a city-state with entrepôt role for limited trade of mainland China to the Western world. As a city-state, Hong Kong experienced fast "modernization" from the 1960s to the 1980s. The inauguration of Kwai Chung container port in 1972 provided Hong Kong with a powerful foundation and efficient gateway for its export-based labor-intensive economy. This former colony has benefited from the dramatic change in the mainland since the 1980s when Dong Xiaoping reformed China's economy by an open-door policy, bringing the country back to the world market economy.

From the late 1970s to the mid-1990s, Kwai Chung Port saw the change in the role of Hong Kong from serving as a platform to a gateway for the Pearl River Delta in South China. Meanwhile, Hong Kong's export-oriented manufacturing sector expanded and gradually relocated northward to the mainland cities of Shenzhen and of Dongguan, establishing a world-class industrial cluster for manufacturing goods, electronic products, and toys. After about 10 years of extra containerized cargo volume generated from the fast growth of the mainland-Hong Kong re-exports, the port business expanded northward. In the early 1990s, the Shenzhen city government started to collaborate with Hutchison Whampoa from Hong Kong to construct and to operate the Yantian Port. China Merchants Group and Hong Kong Modern Terminals Limited collaborated in building the Shenzhen western port area with Chiwan Port at Shekou as its center. Subsequently, Shenzhen Port, which is composed of the above two port areas, developed at an unprecedented pace. Shenzhen Port took only 15 years to attain the same annual container throughput that Hong Kong spent 30 years to achieve. In other words, Shenzhen successfully built a port of the same size as Kwai Chung Port in Hong Kong within two decades in a place less than 50 km away. More critically, the newly built Shenzhen Port is much closer to the sources of goods manufactured in the Pearl River Delta, not only in a physical sense but also in an institutional sense. The land-transported cargo to Shenzhen Port is not involved in any cross-border checking delay, nor does it require expensive Hong Kong truck drivers. As a result, the monopoly of Hong Kong Port in terms of overseas connections has been completely challenged by Shenzhen, which is a neighboring port operated and nurtured by its own operators.

In examining the transformation of Hong Kong's economic structure over the past three decades, the role of the port has likewise changed. From 1975 to 2000, the proportion of employees in different industries has significantly changed (Figure 3.5). The first half of the 1980s was a prime period for Hong Kong's manufacturing industry. Meanwhile, mainland China had already opened up to the outside world, and the manufacturing industry in Hong Kong had not yet moved north. After 1986, the economy of the Shenzhen Special Economic Zone started to grow, and many manufacturing companies were established in the Pearl River

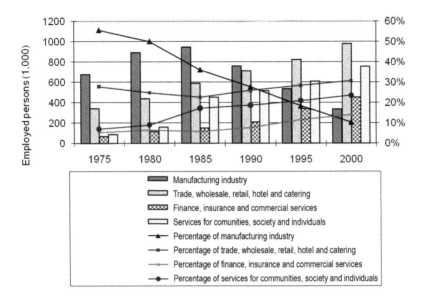

Figure 3.5   Employed Persons and Employment Rate from 1975 to 2000 in Hong Kong

Source: Census and Statistics Department, Government of Hong Kong Special Administrative Region

Delta, with a huge agglomeration of export-oriented factories in Shenzhen and Dongguan. Consequently, the region's total trade volume increased dramatically, thus significantly developing Hong Kong's port industry. Therefore, the last two decades of the twentieth century witnessed the most rapid development of the Hong Kong port. However, the northward movement of the export and the manufacturing industries in Hong Kong was followed by the northward movement of the port business itself. The first call of an international container shipping line at the Yantian port in Shenzhen was in 1994, and its container throughput reached half of that of the Hong Kong Port in 2004. Therefore, approximately one third of the total shipping volume in the Pearl River Delta was accomplished by the up-and-coming Shenzhen Port.

In Hong Kong, among land, sea, and air transport, the contributions of maritime shipping to local trade began to decrease in 2000, which can be attributed to two reasons. First, re-exports in Hong Kong comprised more than 95 percent of the city-state's external trade value. Although the total re-exports grew dramatically in the first decade of the twenty-first century, the percentage of the re-exports handled by sea transport decreased from 52 percent in 2001 to 29 percent in 2011 (Figure 3.6). By contrast, air transport and land trucking correspondingly increased, bypassing water-borne transport by 34 percent and 37 percent, respectively. Another possible reason is the share of transshipment in the Hong Kong container traffic. Figure 3.7 indicates that while the total laden container throughput constantly increased in the past two decades, except for a dip in 2009, the share of transshipment increased from about 17 percent in 1992 to almost 70 percent in 2011. In other words, if the double counting caused by transshipment is disregarded (that is, the vessel-to-vessel containers are counted twice in Asian ports), the growth rate of laden container throughput in Hong Kong from 2002 to 2011 was actually less than 1.57 percent. More importantly, the transshipment from the Pearl River Delta accounted to about 29 percent of the total transshipment in 2011. The figures reveal that major services offered by the port in Hong Kong shifted substantially from initially supporting the manufacturing sector to supporting the re-export trade as a bridge between mainland China and the rest of the world via land trucking. These services then changed to supporting a transshipment hub for both the Pearl River Delta region and other places in the Asia-Pacific, when the manufacturing industry left Hong Kong. The emergence of the Shenzhen Port, its rapidly improved services, and the northward movement of the export and manufacturing industries in Hong Kong decreased the status of the country as a regional container gateway port.

The status of the port in Hong Kong's economy also changed as its functions change. The added values created by different components of Hong Kong's transportation and logistics industries have resulted in a transformation since 1996. The sub-sectors related to the aviation industry developed rapidly, but the development of those related to sea and to road transport either stagnated or gradually declined (Wang and Chen 2010). Therefore, the prime period of Hong Kong's status as a major seaport has passed. The Hong Kong government has delayed its decision in 2008 to build the tenth container terminal. This decision was a seemingly right decision with the life cycle of the port in Hong Kong to remain unaltered. By contrast, the decision by the Hong Kong SAR government to construct a large-scale cruise terminal in the location of the former Kai Tak Airport was in line with the economic transformation of SAR because more "soft port" facilities were required, which served the people and the tourist industry. For the port-related logistics industry in Hong Kong, the situation is much complicated as this sector, as a critical part of the global supply chain, has been regionalized and "multi-modalized": not only are the port and its related services becoming regionalized to a large scale, but a more integrated multi-modal logistics service hub is also taking shape in Hong Kong.

In other words, given the high cost of land and labor, Hong Kong can no longer strengthen its old economic structure based on the port that mainly handles staple goods. Rather, Hong Kong can gradually upgrade into a regional financial hub and into a supply chain management center. This shift in function is indicated in Figure 3.8 and Figure 3.9. Figure 3.8 shows that the economy of a trade-based port city is premised on its local or nearby manufacturing industries. A port city conducts trading by road and sea transport, establishing itself as a trade and manufacturing center of a wider economic region. As economic globalization and local economic development intensifies, the cash flows brought by financial services and the flows of trade-based labor and high-end products increase. Consequently, flows of information and air transport, which largely represent business activities and high-end trade of more valuable or time-sensitive goods, will have larger room to develop than the economic activities that are mainly reflected by flows of goods in land transport and maritime shipping. At this time, corridors through which low-end goods flow start to shift to other cheaper ports in the same region, and then new ports gradually emerge. This upgrade and spillover result in the emergence of multi-layered gateways of foreign trade (Figure 3.9). This functional shift precisely mirrors the relationship between Shenzhen and Hong Kong.

In terms of scale, Hong Kong is still one of the busiest container ports (ranked third in 2012), due to the scale of the exports and consumption of Pearl River Delta region of some 50 million population, and the growing international transshipments, and the latter makes this hub port of high connectivity in its directly connected destinations and the service frequency to world leading ports (Knowles 2006). As long as the shipping lines may benefit from their interlining and relaying activities in Hong Kong, the hub role may be retained. But to the city itself, the challenge to make best use of this hub port is three-fold. First, how to increase value-added re-export activities (i.e. to open the boxes and work on the cargo inside them) rather than transshipments? Second, how to maintain good air quality while serving so many transshipment ships which need not come to Hong Kong otherwise? Last but not least, how to improve the quality of the hub by bettering the supportive logistics and other related services which may mutually benefit both the port users and the local logistics sector? As illustrated by Figure 3.10, the value-added activities such as container transshipments taken by global supply chains (GSC) may need four types of support: 1) inter-modal infrastructure and its operation, 2) trade facilitation procedures, 3) institutional support such as regulations, standards and law, and a cluster of local logistics that help and gain from the GSC hubbing activities.

3.6   Conclusion

This chapter discusses the economic relationship between a port and its host city. The economic relationship between the two is premised on their functional relationship. If the port mainly provides enterprises for the host city, a high level of mutual economic dependence exists between them. We divide port cities into four types according to port function. Each type has its own characteristics of development. The economic structure of a port city may change with time. When a port city undergoes national or regional industrialization, the number of enterprises that generate demand for better operations of the port may increase. By contrast, when the city heads into a postmodern economy, its total demand for shipping may decrease. The port and its related service industry may also gradually become a regional logistical pivot place, which will result in the city providing services to the port, and the port may work together with various other gateways such as

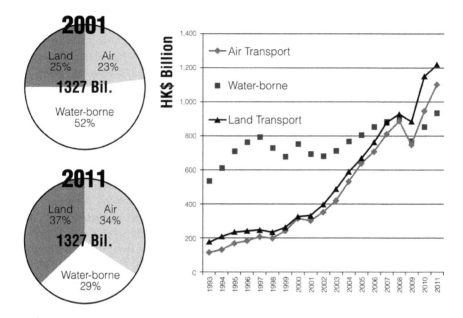

Figure 3.6   Hong Kong Re-exports by Major Modes of Transportation

Source: Hong Kong External Trade Tables 068–070, Census and Statistics Department, Hong Kong SAR Government

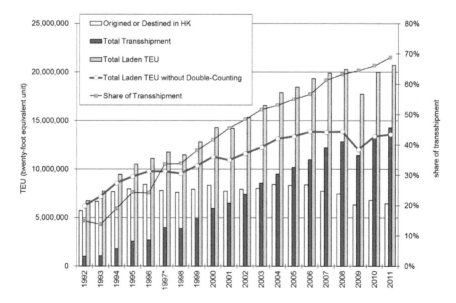

Figure 3.7   Laden Container Throughput and Shares of Laden Container Transshipment, Hong Kong, 1992 to 2011

Source: Department of Census and Statistics, Hong Kong SAR Government

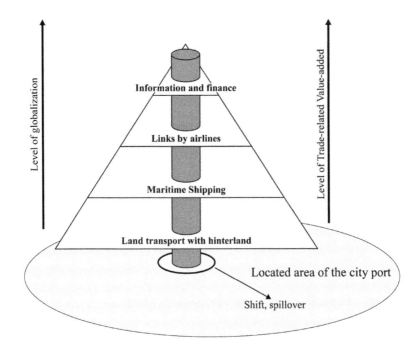

Figure 3.8   Growth Stage Model of Gateway Port Cities (1)

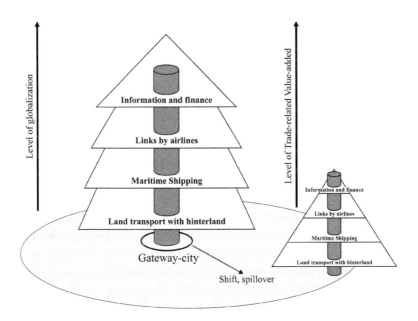

Figure 3.9   Growth Stage Model of Gateway Port Cities (2)

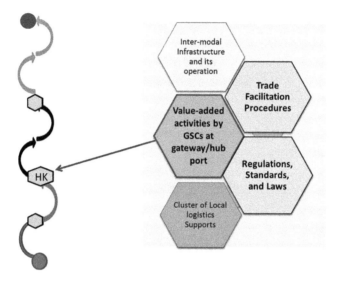

Figure 3.10   Hub-Gateway Port with its Supportive Activities

airports and information hubs. These three functional changes directly influence the dependence of the city on the port. Meanwhile, from the perspective of the port, its functions may change as other conditions change. Consequently, industries based on specific port functions, such as the logistics industry, will implement self-adjustments in terms of their scale and services. These adjustments affect the quantitative or the qualitative transformation of the holistic port-city economic relationship. These transformations have internal causes, such as the different life cycles of ports and cities.

The functional and the scale relationship between the port and the city should be discussed in further examination of the port-city economic relationship. The scale relationship involves two aspects, namely, economic and spatial scales. The economic scale refers to the extent of contributions of the port to its host city. This aspect is discussed in this chapter. The port-city spatial relationship is important for the planning and development of a port city and has a lasting effect. This aspect is explored in the next chapter.

Chapter 4
Port-City Spatial Relations

4.1 Introduction: The Port-City Spatial and Scale Relations

A wharf may cover an area of only a few hundred square meters, whereas a port can be as large as over 100 square kilometers. Some ports may even be larger than all the remaining parts of their host cities, thus truly becoming city-ports. From a port-city perspective, therefore, the spatial form and size of the port may have a profound, dynamic, and long-term impact on the morphology of a city. In the fast developing China, this port-city relation has always been interpreted as a positive dynamic in every dimension it may involve. Many Chinese articles share a common idea, that is, "'Building ports to prosper the city, developing the city through its port, utilizing the port for the development of its host city, prospering the port through the city, the port and the city promote each other's growth, and the port and its host city shall share prosperity and decline'. This is a common rule for the development and evolvement of port cities in the world" (Huo 2009). This statement implies that a positive correlation exists between a port's growth and its host city's evolution, either toward prosperity or decline. Given that China is in its fast growing economic stage, getting more shipping activities must be positive to the host city, that is, when a city grows larger in terms of economy and geography, the port business grows naturally; however, such assumption is not the truth. Whether in China or in other countries, a port's prosperity does not necessarily lead to the boom in its host city, and the decline of a port does not necessarily result in the fall of its host city. A port's evolution may be highly correlated with its host city, but things are not always as such. It is not a fixed rule. For instance, several ports, such as Liverpool, declined and their host cities also fell. Meanwhile, some other ports, such as Venice, declined but their host cities made a successful transformation and continued prospering. In addition, a few ports, such as London and New York (Hong Kong may be the next example), once as the largest ports in their regions, have become the most important financial hub for their nations. Indeed, it is possible to while ports stop developing, their host cities may continue to develop other related industries. Furthermore, the scales of some cities, such as La Havre in France and Qinghuangdao in China, remain relatively small while their ports are substantially expanding. On the contrary, cities, such as Wuhan, have always been large, but their ports are relatively small.

Numerous studies in China investigated the interactive relationships between ports and their host cities before or right after its economic reforms started in the later 1970s (see an overview by Wu and Gao 1989). However, at that time, the data were not adequate for a systematic overview. Much later outside China, in his

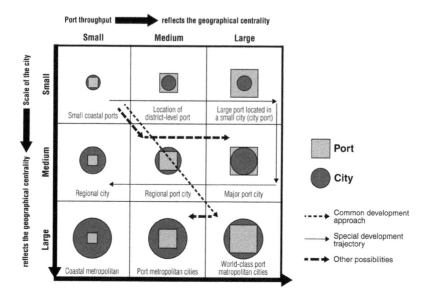

Figure 4.1   Port-City Scale Relationship

Source: Modified from Ducruet and Lee (2006)

doctoral thesis, the French scholar Ducruet pointed out that the scale relationships between ports and cities are diverse (Ducruet and Lee 2006) (Figure 4.1), such that it is possible that a large port is in a small city or a small port is in a large city. A port and its host city do not necessarily have a positive correlation in terms of their scales. This condition can be explained by at least two reasons. First, the port-city functional relationships may change. Changes in the scope of the area the port serves (the hinterland narrows down to the port's local area or experiences an international expansion), the market in the forelands (the demand for products produced by the hinterland changes), shipping arrangements (transformation from direct shipment to transit shipment via a key port or vice versa), and maritime technology, may all result in changes in a port's reliance on its host city. Second, when cities start to prosper from functioning, ports may lose their port-based functions because of increasingly expensive lands and labor brought about by rapid urbanization and more convenient transportation. If favorable conditions are available, some of these cities can experience a successful economic overhaul, and their scales (i.e. population and GDP) may expand even further regardless of their declined functions as ports.

Nevertheless, the port-city scale relationship is a not comprehensive spatial description of the relationship between a port and its host city. An in-depth examination of the port-city spatial relationship should include the following three aspects: (a) the spatial forms of the port and its host city as well as their evolution; (b) the relationship between the port and the distribution of port-based

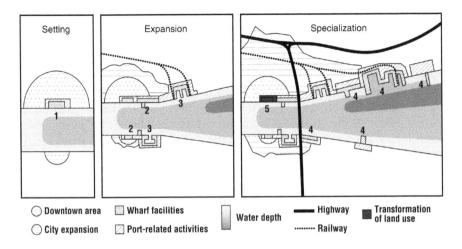

Figure 4.2   **Classic Evolution of a River Port-City Relationship: 'Anyport' Model**

Source: The Geography of Transport Systems, Rodrigue, J.-P., Comtois, C., Slack, B. 2009 New York: Routledge. Reproduced by permission of Taylor & Francis Books UK

and logistical industries; and (c) the opportunities and challenges resulting from the spatial development at various scales.

4.2   The Spatial Form of the Port-City and its Evolution

The spatial form or morphology of a port in its host city and the port's evolution is a dynamic process. Meanwhile, the functions of different port districts and their roles in the host city are also changing. This condition is well elaborated by British scholar, J. Bird, who developed the Anyport Model based on the examination and comparison of the vicissitudes of a large number of ports in Britain (Bird 1963). This model aimed to describe the development of a typical port-city spatial relationship in the process of the port's evolution. After a few modifications and revision later by others, the model now is widely recognized as one of few models in transport geography that have been validated in many places (Figure 4.2). The process of transformation, where a small fishing port transforms to one that can handle different kinds of cargos with specialized terminals, can be described in six stages (Bird 1963; 1971). The first stage involves the initial setting of the fishing port; its host city is just a small town. The second stage witnesses the initial expansion of the fishing port. As the host city expands, a bridge is built and docks are constructed on the other side of the river. Passengers and cargos can then be ferried. In the third stage, ships become larger and coastwise quays can no longer satisfy the increasing demand for cargo shipment. Consequently, piers and excavated-in basins are constructed. Meanwhile, the host city continues expanding and gaps in land rent widen. Residential, industrial, and commercial areas are separated. The fourth stage sees the port's further expansion to deeper waters and the emergence of specially-planned port areas and transport facilities for the port's exclusive use. This development indicates that the port's operation starts to march toward economies of scale. Furthermore, the newly-established port district is further separated from the city center where other main urban functions are clustered. The port in the fifth stage has larger-scale port districts with specially designed railways; in addition, the hinterland considerably widens and the wharf is more specialized. At the same time, the scale of the host city keeps expanding, but in the opposite direction, that is, the port expands toward deeper waters in the lower stream of the river whereas the city expands toward the upper reaches. In the final stage, the port districts are completely specialized. Rodrigue et al. (2009) suggested that the original port district adjacent to the city becomes obsolete as these special wharfs for containers and bulk cargos are built.

Several points are worth noting when applying the Anyport Model to China's present conditions. First, the spatial evolution process of a port described in the model is still valid. However, in considering China's rapid urbanization and large-scale initiation of many port constructions, it is unlikely that any successful ports have followed this model stage by stage. Second, China's coastal cities had been closed to the outside for three decades since the establishment of the People's Republic in 1949. This was especially true in the case of Fujian Province, because of the political and military tension between the two sides of the Taiwan Strait. When many ports in China started to operate after 1978, international maritime transport already entered the era of containerization and ship enlargement. This situation is different from those that occurred in most European and American countries. Third, the "city" described in this model refers to a rather small-scale unit compared with current Chinese cities. As a result, a more complicated model should be developed when examining the port-city relations in China. In fact, several cities, such as Shenzhen, have more than one port in its jurisdiction. It is worth of reiterating that, mentioned earlier in Chapters 1 and 2, unlike the conceptual models provided by Hall and Jacobs (2012) where port-city spatial dynamics are fundamentally analyzed functionally, the administrative jurisdictions in China play a big role in locating and expending ports, since a spatial expansion of ports should be regarded as part of the evolution of *the routines of the states* rather than *the routines of the firms* here in this country. Fourth, this model gives a detailed description of the evolution of transportation in the ports and nearby areas as well as the transformation of wharfs. However, at present, ports handle either diversified or specialized cargos, which may not share common features with ports described in the Anyport Model.

Therefore, a four tri-dimensional models of the port-city spatial relationships is developed here to extend the coverage of the Anyport Model. These models describe the different types of transformations of the port-city spatial relationships found in China today (Figures 4.3–4.7). Time span, distance from the center, and space taken by the port are the three dimensions in each model.

The first model (A) shown in Figure 4.3 is called the "Gradual Change model". This kind of port gradually becomes far apart from the city center in the process of development, either because the port shifts to other places or because the city moves away from it. When the space taken by the port grows to a certain scale, as the port's operation becomes more efficient or the demand for ports reduces, some wharfs or port districts are converted into lands for other use, such as waterfront redevelopment. Consequently, the port's total land use reduces. This model can be regarded as an alternative description of the Anyport Model (see Case study 4.1). The second model (B) is called the "Distant Start-up model" (Figure 4.4). This kind of port mainly consists of seaports such as Port of Tanjung Pelepas in Malaysia. Such large-scale ports are initially located at a suitable distance from the host cities. They are originally planned to anchor larger ships and have specialized functions that differ from those of their host cities. Therefore, they do not have relationships with local or small fishing ports. In addition, these ports may be specially built for national or regional strategic considerations; as a result, their host cities may be much smaller than the ports themselves. Among these ports are the rebuilt port of La Havre in France after the Second World War and that of Fangchenggang in China, which was built for the Vietnam War (see Case study 4.2). These ports are not closer to the city center and their scales are larger due to increased throughput. Building this kind of port requires a large amount of capital investment at the beginning; nevertheless, they may not need much input in a later time for relocation.

Model C is the opposite of Model A and is called the "Spatial Leap-forward model". In this model, a port leaps forward to another location distant from the city center due to limited space for expansion, a shortage of deep waters, or a combination of the two (e.g. Marseille in France and Tianjin and Ningbo in China) (see Case study 4.3). This kind of spatial leap-forward results in a longer distance from the port to its host city and usually leads to an increased amount of land for port use. Meanwhile, the port that leaped forward can benefit from cheaper land price and, consequently, to often larger spaces as catchment area for port-related logistics activities. This situation would not be in conflict with the future development of the city, especially the city center's further expansion. The original port districts gradually become obsolete, the previous waterfront facilities abandoned, and land use changed.

The fourth model (D) is called the "Dual port model", and is a variation of Model C. In this model, both the old and the new ports co-exist in the same host city and enjoy continued expansion simultaneously, which is different from the condition in the Spatial Leap-forward model, in which the new port rises while the old one falls. As a result, the space taken by the ports continues to enlarge, and each port takes on separate functions in specialized areas. If we focus beyond a single city, Model D evidently has its own variation caused by administrative division and different governance systems, that is, two ports (owned by the same proprietor) are close to each other but are located in two different cities. This situation is similar with that in Model D (i.e., two ports co-exist in the same city) if the administrative boundary between the cities is not taken into consideration. However, when taking into account inter-city competitions and different land management, the land price of the far-side port may not be as cheap as that in Model C, because the "far" port may not be far from the center of the other city, leading to more uncertainties in urban land use.

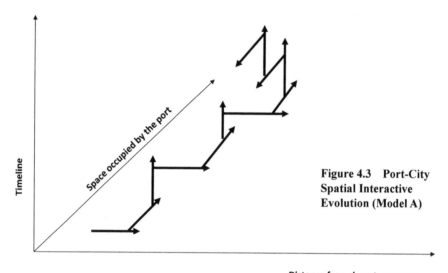

Figure 4.3   Port-City Spatial Interactive Evolution (Model A)

Distance from downtown area

Model A: Port districts gradually move and expand outward from the city center. New port districts expand faster than the gradual abandonment of old port districts; as a result, the total land area occupied by the port gradually widens as time goes by, especially in areas that become close to deep waters and away from the city center. However, because old port districts are used for other purposes, the amount of land occupied by the port starts to decrease. This process is, in fact, a tri-dimensional description of the Anyport Model.

Case study 4.1   Guangzhou

Biographies of Usurers in *Records of the Grand Historian* recorded 23 large cities in the Qin and Han Dynasties (200 BC to 90 AD). Only one port city existed among these cities – the current Panyu in Guangzhou Province (Huang, 1951: 23). Numerous studies agree that Panyu has become one of the key ports for foreign trade in China (Zhao, 2006: 120). Wharfs recorded by historical books had appeared in Guangzhou since the Sui Dynasty; one was located at Poshan (presently Huifu Road West), and the other was in Xilaichudi (currently Hualin Temple). Guangzhou was the biggest port city in China during the Tang Dynasty.

This port was divided into two, namely, an inner port and an outer port. The latter included today's Tuen Muen in Hong Kong and Nanhai Temple in Huangpugang South, whereas the former included Guangta Port (presently Guangta Temple on Guangta Road) and Lanhu Port (currently Liuhua Lake). During the Song Dynasty, the outer port included Fuxu Town, Pazhou Wharf, and Datong Port (near the present Huadi), whereas the inner port included Xi'ao (currently Nanhao Street) and Donghao (today's Qingshuihao Street). During the Qing Dynasty, Huangpu became the outer port of Guangzhou Port. In 1817, the import value through Huangpu amounted to more than 19.71 million silver dollars, accounting for more than 80% of the total import value of Guangdong Province (23.48 million). These data indicated that Huangpu was a key gateway for most of the import and export activities in Guangdong Province. The port district moved east after the establishment of the Republic of China in 1912. In 1937, the formal construction of Huangpu Port commenced. After the establishment of the People's Republic of China in 1949, the original Huangpu Port was reconstructed due to insufficient space. The newly constructed port gradually moved southeast to Shiziyang. Then, Huangpu New Port, Xinsha Port, and Nansha Port were built in the 1950s, 1980s and early 21st century, respectively. These constructions witnessed the gradual outward movement of Guangzhou Port.

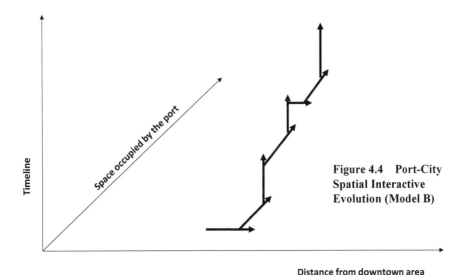

Figure 4.4   Port-City Spatial Interactive Evolution (Model B)

Model B: The port is initially constructed at a suitable distance from the city center. The term "suitable" means that, on the one hand, the port may have its non-physical support, such as insurance and banking services for the shippers and shipping companies being located in the city conveniently; on the other hand, the port expansion would not affect the city's own expansion and vice versa. Spatially, when the port becomes larger in scale, it tends to develop further away from the city. One of the important reasons for the emergence of these kinds of ports is that they are primarily designed as regional transshipment stations or gateway ports. Such ports have sufficient seaward space for expansion by design, and their negative influences on the functions of their host cities are limited.

Case study 4.2   Zhanjiang Port

Zhanjiang was originally called "Kwangchowan." In 1897, French warship *Bayard* found Kwangchowan as an excellent deep-water port when seeking shelter from a typhoon. Then, the officers of the ship submitted a written statement to the French government to rent Kwangchowan as a French port, raising the curtain of Zhanjiang's colonial history. The Chinese and French governments signed the *Sino-French Treaty of Kwangchowan Concession* on the 16th of November, 1899, and Kwangchowan was leased to France for 99 years. The port was taken by the Japanese in 1943 and was returned to China after the Japanese capitulated on the 18th of August, 1945. In the 1950s, Zhanjiang became the largest port established by the People's Republic of China. This port city is not only endowed with a large land area, but also has the highest shoreline-area ratio among all other ports in the country. This port gradually expanded toward the inland areas for several reasons, but it expanded seaward during the three decades since 1978. The newly built iron ore wharfs for Baoshan Iron-Steel Co. gradually entered Dongxing Island, thus moving away from the current urban areas. This is a typical example of Model B.

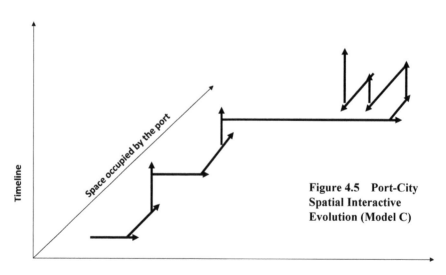

Figure 4.5   Port-City Spatial Interactive Evolution (Model C)

Model C: This exemplifies ports that have spatially leaped forward. At the beginning, the port develops near its host city, thus showing interactive growth. However, the port leaps to another location, dozens of kilometers away from the host city, in order to overcome the technical (i.e., insufficient depth of water due to larger ships) and spatial (i.e., insufficient land or high land rent) constraints at a certain time point. The original port district near the city declines gradually, while the new site expands.

Case study 4.3   Tianjin

Tianjin is one of the cities with a long history, along with Shanghai, Guangzhou, Ningbo, Quanzhou, and so on. Since the Tang Dynasty, Tianjian, consisting of Zhigu and Haijin at that time, has become the pivotal hub of water transport of grain to the capital in the north section of the Grand Canal owing to its location at the joint point of the Haihe River and the Grand Canal. The Ming government established the Tianjin administrative district in 1404, which began to function as the "Gateway for the capital city and its environs" (Fan, 2004). Since then, inland water transport and wharf-related business in Tianjin prospered for several hundred years. After Tianjin's colonization in 1860, Britain, France, and the US established their own concessions at Zizhulin alongside the Haihe River. To develop sea transport in Tianjin, Western countries built the Zizhulin concession wharf alongside the river. Although Tianjin was equipped with wharfs and concessions, it only functioned as a sub-line port in terms of internal and international trade. International trade was mainly handled through Shanghai Port as a transshipment station. Since 1904, Tianjin's direct foreign trade has gradually increased, making it the largest estuary port in north China for a long time. After the first wharf for sea transport in Tianjin New Harbor at Tanggu started to operate, it entered the era of a seaport city.

Then, after the establishment of the People's Republic of China, water transport in the Haihe River began to decline. Bridges and facilities preventing sea water encroachment over the Haihe River all rendered Tianjin into an insignificant estuary port. What remained were small and medium-scale water transport activities. Tianjin New Harbor became the first port with specialized container terminals in China after the reform and opening up in 1978. Geographically speaking, container terminals, bonded areas, and Tianjin Economic-technological Development Area (TEDA) altogether comprised the cornerstone of the establishment of Binhai New Area in 2004. In other words, the establishment of Binhai New Area symbolized a new port-city relationship.

According to the abovementioned historical evolution of Tianjin Port, Tianjin's port-city relationship differs greatly from those of several historically famous port cities in China in terms of spatial evolution. First, unlike Guangzhou and Shanghai, Tianjin was originally established for another city (i.e., Beijing). In other words, Tianjin was established as a port that served the hinterland beyond itself; it might be called "a logistical center" in today's setting. This is a very significant and influential point. Second, from the earliest Sanchahe wharf, to Zizhulin wharf, and then Tianjin New Harbor at Tanggu, Tianjin Port experienced the transformation from an estuary port to a seaport at an earlier time. We have especially noticed that this transformation occurred before the containerization of international shipping in the second half of the 20th century.

However, Guangzhou's spatial transformation from an estuary port to the latter Guangzhou New Harbor and the recent Nansha Port, as well as the Shanghai Ports' spatial leap-forward from Pujiang to the mouth of the Yangtze River (Zhanghuabang), the latter Waigaoqiao, and the latest Yangshan, both happened after the containerization of international shipping. Nevertheless, the availability of a containerized port is not substantially related with Tianjin's transformation from a commercial city to an industrial one. In fact, after Tianjin Port moved to Tanggu, it did not immediately participate in international trade until 40 years later, when China reformed and opened up. The industry in Tianjin did not concentrate in Tanggu until TEDA was established.

The establishment of Binhai New Area and the sea reclamation in Tianjin resulted in massive geographical changes. First, the spatial distribution of export-oriented processing and manufacturing industries, heavy and chemical industries, as well as large-scale machinery manufacturing industries in Binhai New Area had no relationship with Tianjin's city center. At the administrative level, these industries brought employment, considerable energy and water consumption, GDP growth, and tax revenue to Tianjin. In Binhai New Area, the port's rapid spatial shift has led to the emergence of a large amount of lands between the city center and the port area, creating spaces in which these industries established operations. In 2009, however, urban planners noticed the absence of a commercial district between the growing numbers of industries and the port. Therefore, they came up with a bolder idea that the port be divided into two parts (north and south) and move each of them northwards and southwards, respectively, leaving some space in between for the development of a commercial district.

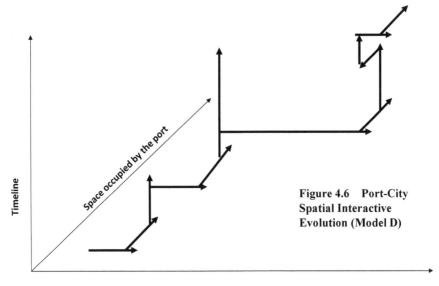

Figure 4.6 Port-City Spatial Interactive Evolution (Model D)

Distance from downtown area

Model D: The port and its host city are interdependent at the beginning. The port grows and expands gradually along the waterfront near the city. However, considering the increasingly developing urban areas, increased competition among different land uses, pressure over environmental protection, higher requirements for water depth, and so on, the port experiences a spatial leap-forward to a distant place from the city center; therefore, a second port emerges as the old one continues its operation.

Case study 4.4   Dalian Port

Dalian became a port city not a long time ago. A commercial port was established in Dalian in 1899, when it was turned into a Russian concession. Railways connecting to the port were built after Japan's takeover in 1905. After nearly ten years, Dalian developed into the largest port in northeast China (Yao, 2005). From then to 1984, when the Dalian Development Zone was established at Dayao Bay, Dalian Port had gradually extended along the coast of Dalian Bay and gradually went further away from the city center. It gradually turned into a large-scale multifunctional port with a number of port districts. Later on, the construction of the Dalian Development Zone and a new port at Dayao Bay marked the first stage of Dalian Port's outward leap-forward. This typical way of establishing "Development Zone + port" and the resulting new urban areas helped transform Dalian into a city with two hubs.

This process took much less time than in some foreign cities, such as Marseille in France; one of the reasons for this is that the majority of workers and residents in the Dalian Development Zone are not local people, thus leading to a rapid increase in the number of permanent residents there. On the contrary, a large amount of commuting population is in the Tianjin Development Zone.

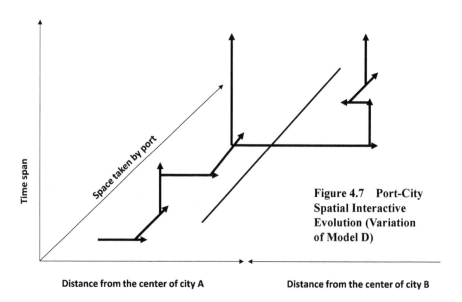

Figure 4.7   Port-City Spatial Interactive Evolution (Variation of Model D)

Distance from the center of city A

Distance from the center of city B

Variation of Model D: When the scope of the dual port in a city extends across the administrative boundary of another nearby city, Model D changes into one in which two ports owned by the same proprietor coexist in two cities. For the port system itself, this situation is not special; however, is the situation is different when viewed from either the administrative or management perspective.

Case study 4.5   A Port in Two Cities—Hong Kong/Shenzhen

Hong Kong built its first specialized container port at Kwai Chung in 1972. Since then, Hong Kong Port has gradually moved from the Victoria Harbor to the current suburban town. At present, Kwai Chung has already been part of Hong Kong's urban area. Since the second half of the 1980s, manufacturing industries in Hong Kong have moved northward to Shenzhen and Dongguan, as well as the whole east coast of the Pearl River Delta.

As a response to this spatial shift, the Yantian and Shekou port districts in Shenzhen, which are located at the east and the west of Hong Kong, respectively, were built in the early 1990s. These two port districts' main shareholders were listed companies from Hong Kong. This situation was different from that of other peer ports in Mainland China. Particularly, 73% of the shares of Yantian Port were owned by Hutchison Port Holdings, the world's biggest port company, and the largest shareholder of Kwai Chung Port in Hong Kong. As a result, the rise of ports in Shenzhen actually reflected Hong Kong Port's gradual departure from its original city center. However, although Shenzhen and Hong Kong are close to each other and are interdependent, port managements and data collections are conducted by different governments under the policy of "one country, two systems." As a result, people often misunderstand that ports in Hong Kong and Shenzhen are different ones and that they are in competition with each other.

In fact, this case is considered very special, given that one port is in two cities with different administrative boundaries (see Figure 7-1 Property Ownership of Major Container Ports in Greater Pearl River Delta). Shenzhen Port (i.e., Yantian and Shekou) is simply a geographical extension of Kwai Chun Port. The Shenzhen Special Economic Zone has prospered and urbanized very rapidly courtesy of its proximity to Hong Kong and favorable government policies since its establishment in 1980. Shenzhen's population was less than 100 thousand in 1980. By 2006, this already increased to more than one million in 2006. Shenzhen's amazingly rapid urbanization resulted in its dramatic expansion. The originally marginalized Shekou and Yantian have become integrated with Shenzhen's urban area, and the population in Shekou and Nanshan has also exceeded one million, making Shenzhen one of the largest port cities in the world.

The rapid expansion of this port in two cities will not have any great influences on Hong Kong's urban layout. However, the evolution of land use in Shenzhen will be heavily influenced by the port's location, its future expansion, and related transportation of large amounts of containers.

The rapid expansion of this port in two cities will not have any great influences on Hong Kong's urban layout. However, the evolution of land use in Shenzhen will be heavily influenced by the port's location, its future expansion, and related transportation of large amounts of containers.

Next, we explore the spatial evolution processes of the abovementioned four types of ports to discover and sum up several inherent laws of ports' development. However, we still consider exceptions. We observe that several ports' development processes deviate from those described above. For example, Shenzhen simultaneously built two port districts (i.e., Yantian and Shekou), both of which are located not far away from the city center. This is a very special case throughout the world. If we take a look at these two port districts from a wider perspective, Shenzhen's case can be regarded as a variation of Model D. In the late 1980s, when port investors from Hong Kong began to build these two port districts, Shenzhen did not have any characteristics that a metropolis should have. Both Yantian and Shekou were relatively far from Shenzhen's city center at that time. However, nobody could have predicted that Shenzhen, a small city with a population of less than 100,000 in 1980, would become a megalopolis with a population of more than 10 million in less than 30 years. Therefore, the port-city spatial relationship in Shenzhen is the most difficult to examine among port cities in China. Such rapid changes in the scale of development remind us of several issues related to the spatial scale in the port-city interplay.

Shenzhen as such a variation has already encountered a serious while unique landuse conflict with its urban development. when the city government decided in 2012 to proceed Qianhai, a new financial district at a reclaimed coastal area, located just one kilometer away from and in beween Shekou and Dachan Bay port districts. The city government has to think of moving a brand new terminal away from its existing location, and/or stop the next phase of port expansion.

4.3   The Spatial Scale of Port Cities

Our study on the spatial relation between a port's spatial expansion and shift and the location of the city center is in line with the general trend of studies that examine the development of port cities in the world. We agree with James Bird's Anyport Model, which argues that a port moves further away from the center of its host city until it becomes separated due to the fact that the growth of the port is in an increasing conflict with the city's land use, transportation, and waterfront resources rather than the location of the central business district of the host city. However, the geographical scale of the port-city spatial interplay is usually neglected by studies in China as well as in other countries. Rescaling in space has become one of the buzzwords in the New Marxist economic geography and is regarded as part of the evolution of the social relations of production in the reproduction process. As producers (e.g., multinational corporations) change the spatial scale of their production networks, the nation-state governments have to actively or passively adjust and even create new scales in jurisdiction and administration; they also need to manage or govern the territorial changes to adapt to new circumstances brought by globalization (Swyngedouw 2004).

This key concept should be applied to studies on the interplays and relations between ports and cities in China, because the concept can help explain why many ports in China experienced spatial leap-forward. In this country, the enlargement of the administrative jurisdictions of host cities is an important prerequisite for the completion of their ports' spatial leap-forward. Guangzhou and Fuzhou are two very convincing examples to illustrate this. In 1995, several scholars suggested that Guangzhou should develop its deepwater port at Nansha, located at the southernmost part of the mouth of the Pearl River, to be better prepared for trade globalization and ship enlargement (Wu et al. 1999). However, Nansha Port was not built until Panyu (a county-level city that did not have enough administrative capital to build its own port but competed with ports in Guangzhou, Shenzhen, and even Hong Kong) was re-incorporated into Guangzhou in 2000. In other words, it was after Panyu's re-incorporation that Guangzhou was able to build Nansha Port, which was beyond its original administrative boundary. During the 20-year period between China's reform and opening up in 1978 and Panyu's re-incorporation into Guangzhou, the furthest port district that Guangzhou could build was only Xinsha Port, located on the boundary of Guangzhou.

Fuzhou is another interesting example. Fuzhou Port started to thrive before Quanzhou Port declined 900 years ago. Fuzhou Port regained its function as a port in the late 1970s, when China started its economic reforms and opened up to the world market. However, Fuzhou Port had been constrained in the Mawei port district and had not been able to become a seaport for more than two decades since then, whereas Xiamen, a port in the same province, developed rapidly. The main reason for this was that no coast with sufficient depth of water could be found at the mouth of Minjing River within Fuzhou's jurisdiction. Fuqing, a county-level city, was incorporated into Fuzhou in 1983. After Fuqing's incorporation, Fuzhou started to consider building the present Fuzhou New Port at Jiangyin Bay in Fuqing.

Notably, the *Overall Urban Planning of Fuzhou (1995–2010)*, which was formally ratified by the State Council, did not consider Fuqing until May 1999. Both Panyu and Fuqing are prosperous county-level cities that have never been included in the Chinese government's port planning. As reported, Jiangyin Bay, the location of Fuzhou Port today, was founded in 1992 by Indonesian-Chinese Wenjing Lin and Fuqing's party secretary Zhixuan Lian, who were looking for coasts with good conditions for the purpose of boosting up Fuqing's industrial development. Instead of waiting for state-funded investigations, port resources in Jiangyin Bay were prospected, assessed, and planned with Lin's own capital investment. After eight years, the central government approved the project to build today's Fuzhou New Port in Fuqing.[1] Later on, Lin sold most of his shares in Jiangyin Port (i.e. currently Fuzhou New Port) to Singaporean Port Group (PSA), one of the top tree container terminal operators in the world; the latter has since become one of the shareholders of Jiangyin Port. The concept of "Greater Fuzhou" has never been proposed until 2006, when the *Strategic Research on Fuzhou's Urban Spatial Development* was published. At that time, the concept of the Fuzhou New Port was introduced, and all the foreign trade shipping lines shifted from the original Mawei Port to Fuzhou New Port at Jiangyin Bay, including French company CMA-CGM, which entered the new port district at a much earlier time than others (Figure 4.8).

---

1 See Fuqing's website: http://www.fuqing.com/content.asp-catid=114151791.htm, and website of overseas Chinese: http://www.chinaqw.com.cn/rwjj/szqs/200803/19/110536.shtml.

Figure 4.8   Fuzhou New Port

4.4   The Relationships Between the Distribution of Urban Industries and the Locations of Ports

With regards the spatial influences of urban industries on the host cities, the early Lorry Model has attempted to differentiate between industries and employments that serve the cities themselves and those that meet the demands outside the cities. Although this kind of differentiation has been criticized for its over-simplicity, it is conceptually a very good framework for our initial analysis, and can be used in our studies on port cities. Indisputably, ports play an important role in the outflows of cargos in any cities that are called port cities. Nevertheless, wharfs and shorelines used for internal communications or commuting, such as ferry terminals that transport people between two river banks, are not taken into consideration when studying the port-city relationships. Therefore, industries sensitive to the locations of the ports are either relevant to the external trade of the host cities (both domestic and abroad) or the ports themselves. Both the internal and external industrial relationships between the ports and their host cities, as well as their interrelationships, can be demonstrated in the four quadrants in Figure 4.9.

Among the industries aiming to serve local needs, if the service recipients are the cities, they should be spatially separated from the ports; if they aim at serving the ports, they should be near the ports, which is not discussed here (Figure 4.9). However, among the industrial sectors and businesses that cater to external needs, which ones are closely related to the ports? Which ones have more

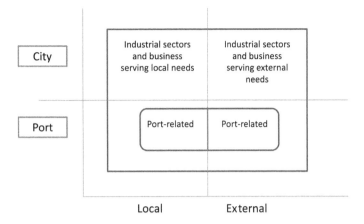

Figure 4.9   Binary Industrial Classifications of Port Cities

stable relationships with the ports? Which ones have interactive relationships with the ports? Which ones need to be more spatially close to the ports? These questions should be answered by a relatively new concept, that is, by modern supply chain management. By referring to concepts in modern supply chain management, several studies believe that the industrial distributions and their changes in different countries within the context of economic globalization should be viewed from the perspective of Global Production Networks (Henderson et al. 2002) or Global Value Chains (Gereffi et al. 2005). For buyer-driven production chains, such as Toys "R" Us, buyers themselves or their agents (i.e. supply chain managers) take the initiative to conduct global sourcing, choose manufacturers to produce specialized products according to their designs and specifications, and organize the entire logistical supply. For the ports, what differentiates the operation from traditional producer-driven productions is that buyers or their supply chain managers consider the place of production, transportation, and the total logistical cost (including the choice of ports for import and export) before products are manufactured. In this model, overseas market and foreland are holding the initiative while the "hinterland" (i.e. the area where products are produced) is being selected; in other words, "where and how many to distribute" determines "where and how many to produce". Once a product is produced – from raw materials to various components – in different places or even in different countries, the final assembly is likely be performed in port cities with low labor and land costs and convenient transportation to ship the products to the global or specialized markets.

When examining the port-city relations in China from this perspective as well as the status quo of economic globalization, both economic and spatial relations between ports and their host cities apparently make adjustments according to changes and demands of the global supply chain (Liu, 2005). If we divide

Table 4.1 Classification of Representative Industries Incorporated in the Global Supply Chains in the Coastal Cities of China (Cities in Brackets are Typical Cities)

	International market-based	Initially international market-oriented and then gradually entered the Chinese market	Equal importance given to international and Chinese market at first	Focus on the Chinese market and then begin to enter the international market	Chinese market-based
External source	Assembly of branded electronic products (Dongguan)		Middle- and high-end electrical products (Shanghai)		**Petrol chemical industry** (Huizhou, Zhuhai, Quanzhou, and Zhanjiang)
External and internal sources	Branded clothing (Dongguan)	Cell phones (Tianjin and Xiamen), electrical components (Shanghai and Suzhou), **automobiles** (Guangzhou), and **ship building** (Qingdao, Dalian, Shanghai, and Zhoushan)			**Iron and steel** (Tangshan and Shanghai)
Internal source	Toys (the Pearl River Delta)	Container and trailer manufacturing (Shenzhen, Qingdao, and Shanghai)	Organic fruits and vegetables (Zhangzhou)	Footwear (Wenzhou and Quanzhou), electrical appliances (Qingdao, Pearl River Delta, and Yangtze River Delta), ceramics (Foshan), and furniture (Shunde)	

Note: the underlined industries require special (non-container) terminals

these supply chains into final consumer goods (mainly containerized products through sea transport) and staple industrial products (mainly bulk cargos such as crude oil and ores, etc.) and consider the hallmarks of China's export-oriented processing resulting from its entry into economic globalization (Table 4.1), the industries and factories concentrated near the ports can be divided into two types according to their relations with the construction and development of the ports (Figure 4.10). The first type of industries and factories has an interactive agglomeration relationship with the ports (mainly container terminals). This kind of relationship is caused by the abovementioned buyer-driven production chains. The interactive agglomeration means that the cluster of these industries near the ports and the expansion of container terminals are both independent and mutually reinforcing. This means that an increase in the concentration of the industries corresponds to an increase in the number of shipping routes and services to be added; similarly, an increase in the number of added shipping services corresponds to a greater attraction of the industries. The second type of industries and factories are constructed simultaneously with the terminals (synchronized development). Several industries and factories are raw-material intensive and are sensitive to transportation costs of semi-final products or raw materials. Thus, such industries typically need dedicated wharfs to reduce the total cost. As a result, the terminals and factories are constructed at the same time in a site where both the inland market/hinterland and the sea/foreland are easily accessible. Hence, these industries and factories do not necessarily get close to public ports, but they are definitely ones that should be located along the seashore and deepwater front (see Figure 4.11 for classification of industries surrounding a port).

4.5   Processing-Trade-Port Zones with the Integration of Production, Trade, and Logistics

Adjustments in international trade environment and policy lead to drastic changes in the transportation system. This kind of influence is fully illustrated in Slack's research on the influences of the formation of the North American Free Trade Area (NAFTA) on various means of transportation in Canada (Slack 1993). In relation to China's case, we pay special attention to the fact that many cities have created special zones at port for buyer-driven industries to adapt to the adjustments of global supply chains. These zones include export processing zones, bonded zones, and bonded port districts. China's government has been formulating the "international articulation space" (Wang and Oliver 2006) in the country since the early 1980s. This articulation space has experienced a process of extension and intensification. Extensively, these zones have gradually spread out to many port cities or even non-port cities. Intensively, as time goes by, newly established zones have become increasingly international in a sense that the business environment tends to be less regulated to attract activities that would otherwise be carried out outside China (Table 4.2).

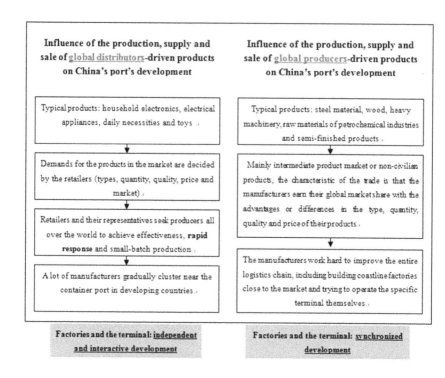

Figure 4.10   Different Production and Sale-Driven Systems after Economic
               Globalization and their Influence on the Cluster of Enterprises near
               the Port and the Terminal's Development

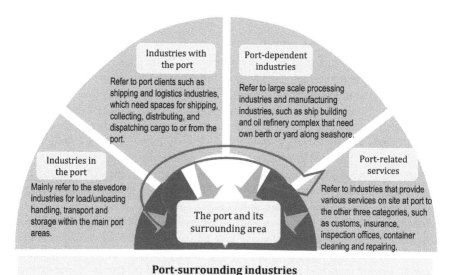

Figure 4.11   Content and Classification of Industries Surrounding a Port

Table 4.2   Classification of Different Kinds of Zones Related to International Port Trade and Relevant Policies

		Bonded port area	Bonded zone	Bonded logistics park	Export processing zone
	Nature	Special zones supervised by the Customs with the following functions: port, logistics, processing, and trade	Special zones supervised by the Customs without the functions of a port	Special zones supervised by the Customs with the functions of a port	Special zones supervised by the Customs without the functions of a port, and are mainly for the export processing business
Customs policies	Customs administration	Administered by the Customs of the People's Republic of China for the Administration of Bonded Port Areas	Administered by the Customs of the People's Republic of China for the Administration of Bonded Zones	Administered by the Customs of the People's Republic of China for the Administration of Bonded Zones	Processing zones can only be set up in economic and technological zones approved by the State Council and administered by customs for the processing zones
	Supervision scope	Bank deposit accounts, contract arrangement, and the standard cost per unit are not implemented. The Customs checks the warehouse according to the import, export, and transit of goods to implement the model of "check in place of review"	Warehouse is supervised by the Customs; in principle, the Customs supervises enterprises	Internet-based, electronic-based, paperless, and intelligent supervision of the zones; in principle, the Customs supervises the zones	The Customs supervises enterprises in the zones and manages the enterprises with electronic records
	Service time	General inspection departments work according to normal working hours; ship inspection division works round-the-clock	Work according to the normal work schedule	Entry and exit are cleared by the Customs round-the-clock	Work according to the normal work schedule
	Import declaration	Get the goods alongside the ship; the goods go directly to the zones; declare the goods with electronic data; prepare for the declaration after getting to the zone	Get the goods in the zones; declare the goods with written declarations; prepare declaration before getting to the zone	Get the goods alongside the ship; the goods go directly to the zones; declare the goods with electronic data; prepare for the declaration after getting to the zone	–

Table 4.2   *Continued*

		Bonded port area	Bonded zone	Bonded logistics park	Export processing zone
Comparison between the port and the zone	Geographical location	The port and the zone are integrated	The area is near the industrial zone; therefore, they have area advantages; it is separated from the port zone by a certain physical distance	Near the port area; therefore, it has several advantages; it is connected to the port through private direct passages	Located in economic and technological zones approved by the State Council and separated by a certain physical distance from the port area
	Correlation	Combination of the functions of bonded zones, bonded logistics, parks and export processing zones	The port and its hinterland are close, the supervised area and the port operate independently	Regional linkage, functional linkage, information linkage and operational linkage	--
	Customs barrier	Automatic unmanned management	Checked by manpower in the barriers	Need manpower to check in the barrier with semi-automatic management	Checked by manpower in the barriers
Foreign exchange policy	Non-trade purchase of foreign exchange	Non-trade purchase of foreign exchange in most areas	Enterprises in bonded zones can only purchase non-trade foreign exchange within certain areas	--	Enterprises in the processing area can buy non-trade foreign exchange after they have submitted the required documents
	Goods flow and capital flow	Goods flow and capital flow can be inconsistent	Goods flow and capital flow can be separated under several terms but with relative restrictions	Goods flow and capital flow can be separated in more areas, which can be trialed in foreign exchange ex payment pilot areas for inconsistent goods flow and capital flow	Goods flow and capital flow must be consistent

Foreign exchange policy	Domestic receipt and payment in foreign exchange	Transactions can be settled in foreign exchange with domestic enterprises	Transactions cannot be settled in foreign exchange with domestic enterprises	Transactions can be settled in foreign exchange with domestic enterprises	The foreign exchange income of the organizations in the zones can be deposited in foreign exchange accounts; all foreign exchange payments can be deducted from the foreign exchange account; no verification or write-off procedures are needed to export foreign exchange receipt and import foreign exchange payment
Tax policies	Export tax rebate	Tax rebate procedures will begin after goods arrive at the zones	Tax rebate procedures will begin after the goods are actually loaded onto the ships	Tax rebate procedures will begin after goods arrive at the zones	Tax rebate is awarded for machines, equipment, raw materials, spare parts, packaging materials, and infrastructure materials that are made in China and are brought to the processing zones for enterprises from the outside
	Turnover tax	No need to declare to the Customs or pay taxes for the circulation of goods among the enterprises in the zones	In cases when levying (or not levying) the turnover tax on products or not is unclear, normal taxation regulations shall apply	Value-added tax and consumption tax are exempted for goods transferred between enterprises in the same park	In cases when levying (or not levying) the turnover tax on products or not is unclear, normal taxation regulations shall apply

Table 4.2 *Concluded*

		Bonded port area	Bonded zone	Bonded logistics park	Export processing zone
Tax policies	Tax refund for special materials	Tax can be refunded for equipment, packaging materials, and reasonable amounts of basic materials for infrastructure made in China and used by enterprises in the bonded zones	Tax cannot be refunded for equipment, packaging materials, and a reasonable amount of basic materials for infrastructure and office buildings for administrative departments made in China and used by enterprises in the bonded zones	Tax can be refunded for equipment, packaging materials, and a reasonable amount of basic materials for infrastructure made in China and used by enterprises in the parks	Tax privileges are given for a reasonable amount of basic construction materials (including water, electricity, and gas) needed by processing enterprises and production and office buildings of administrative departments
Business operation **Business operation**	Re-import business	If one-day-tour goods return to the domestic market, the issue can be handled very easily	If one-day-tour goods return to the domestic market, the issue cannot be handled very easily	If one-day-tour goods return to the domestic market, the issue can be handled very easily	If one-day-tour goods return to the domestic market, some of the processing zones approved by the State Council with the function of bonded logistics can handle the issue very easily
	Distribution business	Enterprises in the bonded zones can distribute their goods by batch with centralized declaration	Goods cannot be distributed outside the zones by batch with guarantee	With guarantee, goods can be distributed by batch with centralized declaration	For enterprises going out of the zones and enterprises entering the zones, the carrying-forward procedure by way of "distributing goods by batch with centralized declaration" can be utilized
	Transit LCL business	Once the goods enter the port areas, the containers can be opened and different goods can be placed into the same container to be shipped on other international shipping liners to a third country or put in transit to other domestic ports	It is difficult to conduct LCL business for transit containers, because the operations are complicated and the cost is relatively high	Transit containers can be opened and different goods placed into them, changing the situation, where only full containers can come and go from the port area	--

Business operation	Export container goods	For domestic products brought into the logistics storage park in the zones, comprehensive handling, such as export container shipping or simple commercial processing, can be applied to distribute the products worldwide; for imported bonded products brought into the zones, comprehensive handling, such as export container shipping or simple commercial processing, can be applied to distribute the products returned worldwide; for the products returned to the domestic market, import procedures need to be implemented according to regulations. The Customs is the same, the returned products can go directly through the zone and only one export declaration is needed	Administered by two Customs agencies; customs transit business; declaration by separate bills	Customs is the same; the returned products can go directly through the zone and only one export declaration is required	Several bonded processing zones with logistics functions and the approval of the State Council can conduct export container shipping procedures with domestic purchase

Sources: Interim Provisions of the State Council of the People's Republic of China on Preferences for the Construction of Ports and Piers with Chinese and Foreign Joint Investment (State Council 1985); Notice of Transmission of the General Office of the State Council on Opinions of Departments like the Ministry of Transportation about Enhancing System Reform of the Department Directly under the Central Government and Dual Leadership of Port Management (Ministry of Transportation 2001); Questions-answers of Director Li Lanxue in the Press Conference about Yangpu Bonded Port Area Convened by Hainan Provincial Government (Website of Haikou Customs of the People's Republic of China 2007); Approval Criteria and Procedures for Establishment of Export Processing Zones (General Administration of Customs et al. 2004)

Model of "zone-port integration" in Tianjin Binbai area

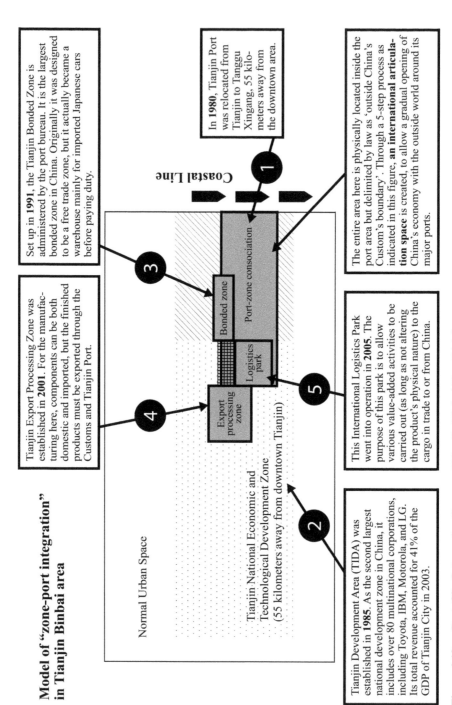

In **1980**, Tianjin Port was relocated from Tianjin to Tanggu Xingang, 55 kilometers away from the downtown area.

The entire area here is physically located inside the port area but delimited by law as 'outside China's Custom's boundary'. Through a 5-step process as indicated in this figure, **an international articulation space** is created, to allow a gradual opening of China's economy with the outside world around its major ports.

Set up in **1991**, the Tianjin Bonded Zone is administered by the port bureau. It is the largest bonded zone in China. Originally it was designed to be a free trade zone, but it actually became a warehouse mainly for imported Japanese cars before paying duty.

Tianjin Export Processing Zone was established in **2001**. For the manufacturing here, components can be both domestic and imported, but the finished products must be exported through the Customs and Tianjin Port.

This International Logistics Park went into operation in **2005**. The purpose of this park is to allow various value-added activities to be carried out (as long as not altering the product's physical nature) to the cargo in trade to or from China.

Tianjin Development Area (TIDA) was established in **1985**. As the second largest national development zone in China, it includes over 80 multinational corporations, including Toyota, IBM, Motorola, and LG. Its total revenue accounted for 41% of the GDP of Tianjin City in 2003.

Coastal Line

Normal Urban Space

Port-zone consociation

Bonded zone

Logistics park

Export processing zone

Tianjin National Economic and Technological Development Zone (55 kilometers away from downtown Tianjin)

Figure 4.12   Port-City Integration Process of Tianjin Binhai New District

These "special zones", ports, and the Customs have very close relations. The efficiency of these special zones is directly related to how well these relations are dealt with. The case of Tianjin Port shows a relatively successful handling of these relations.[2] I have taken Tianjin Port as an example to elaborate the integration process of ports, bonded zones, and development zones (Wang and Olivier 2007). Figure 4.12 demonstrates the chronological evolution of the zone-port interplays since 1980, when Tianjin Port leap-forwarded to its new location in Tanggu (presently New Coastal District). The spatial relations among Tianjin's development zone, bonded zone and container terminals, as well as sites supervised by the Customs, are relatively favorable to the efficient function of the bonded zone. This means that the land of Tianjin City has been divided in a logic from the normal space for inland industries with domestic sources and market, to a special space for industries that use imported raw material or components to produce products for China market, and then to the export processing space and the container terminals that play the role of an interface of foreign trade.

Actually, this kind of zone-port integration is in line with the State Council's general thinking on the gradual development of free trade zones through policy evolution several years ago. Cheng Siwei, the former vice chairman of the Standing Committee of the Chinese National People's Congress, once pointed out that China may complete its transition from the current bonded zones to real free trade zones in a period of ten years (Cheng 2003). "Zone-port consociation", which was initiated in China's major ports in 2003, raised the curtain of this transition. However, many practical issues are worth discussing when this transition is viewed from any aspect of the port itself, the host city, or the whole country.

At the port level, linking up bonded zones to the port districts and creating a truly "free port" is a common problem. Liu Yan (2004) and Dou Ping (2006) conducted detailed and outstanding discussions about how the bonded zone in Shanghai Port can develop into a free trade zone in their respective Master's theses. Both studies have noticed that the means through which efficient and smooth operations are maintained between the bonded zone and the port, which are "inside the territory but outside the Customs", serve as the key factors in such development. According to Dou Ping, one can refer to Chapter D2 about "Free Zones" in Specific Annexes of the *Kyoto Convention* (*International Convention on the Simplification and Harmonization of Customs Procedures*) signed by the World Customs Organization Council in Tokyo on the 18th of May, 1973 for the definition of the phrase "inside the territory but outside the Customs". International open zones, such as free ports, free trade zones, and so on, are called "Free Zones" in this Annex. A "free zone" means a part of the territory of a Contracting Party where any goods introduced are generally regarded, insofar as import duties and taxes are concerned, as being outside the Customs territory,

---

2   The author believes that Tianjin's case is worth learning, but is only limited to the handling of the spatial relationships. At the administrative and governance level, the handling of the relationships is problematic. This issue will be further discussed in Chapter 6.

which is entitled to exemption from usual Customs supervision and control. According to this definition, Dou Ping believes that a "free zone" in a country has three essential characteristics: (a) "Part of the territory", that is, "inside the territory"; (b) "Being outside the Customs Territory", or "outside the Customs"; and (c) "Exemption from usual Customs supervision and control". A zone can be a "free zone" only if these three conditions are satisfied simultaneously (Dou 2006). However, a research on bonded zones in China conducted by Liu Yan (2004) found that in many cases, geographical superiority of these zones (being near the ports) did not bring about economic advantages that they should have enjoyed. This is because the bonded zones and the ports were either "located in the same city but far apart" or were "geographically linked together but poorly coordinated". This problem is mainly caused by the system – at the administrative level, the principle of "separating enterprises from government management" is not truly carried out in many aspects, resulting in different interest groups in the bonded zones and the ports. Consequently, "zone-port integration", which aims at "integrated planning, integrated management, integrated mechanism and integrated policy", cannot be realized because of the absence of the latter two objectives (i.e. integrated mechanism and integrated policy).

At the city level, many ports have spatially leaped forward, leading to the shifting of container terminals, development zones, and bonded zones to a place that is distant from the city center, as in the case of Tianjin. Apart from the improved conditions of the depth of water, the most important reason for this condition is a large amount of land supply. However, analyses of the land leasing prices in development zones are rarely made public. Almost every development zone utilizes very low land rent and long leasing contracts (i.e. from 25 to 50 years) to attract foreign investments. At the same time, the business tax exemption and/or deduction from local municipal governments may be counted as important parts of major revenues of those firms located in the zones. Normally, no significant differences or advantages exist in the land leasing and other incentives among different development zones in the same city; what matters most is their spatial relations with the ports. For example, an analysis has been conducted about the economic benefits of six national-level development zones and ten city-level development zones (both types include bonded zones and export processing zones) in Shanghai (Liu 2003) (see Table 4.3 for development zones at all levels in Shanghai). After using clustering analysis to examine several indicators of these developments zones, Liu divided these into four ranks according to their economic benefits. The first rank has only one development zone, the Jinqiao Export Processing Zone. Similar with the first, the second rank also has just one zone, the Shanghai Waigaoqiao Bonded Zone. The third and the fourth ranks both include seven zones each, with the third rank headed by the Zhangjiang Hi-tech Park. When marking these four ranks of development zones on a map, a negative correlation can be easily identified between their rankings and their distances from Waigaoqiao Container Terminal in Shanghai, that is, development zones closer to the container terminal tend to have better economic benefits.

Table 4.3   Development Zones at All Levels in Shanghai (64 in Total)

National Development Zones

- Minhang Economic and Technological Development Zone
- Lujiazui Finance and Trade Zone
- Songjiang Export Processing Zone
- Qingpu Export Processing Zone
- Shanghai Yangshan Bonded Port Area

- Hongqiao Economic and Technological Development Zone
- Shanghai Waigaoqiao Bonded Zone
- Jinqiao Export Processing Zone
- Minhang Export Processing Zone
- Shanghai Lingang Heavy Equipment Industrial Zone

- Caohejing New Hi-tech Development Zone
- Zhangjiang Hi-tech Park
- Caohejing Export Processing Zone
- Shanghai Lingang Industrial Zone

Municipal Development Zones

- Shanghai Jiading Industrial Zone
- Shanghai Pudong Xinghuo Development Zone
- Shanghai Xinzhuang Industrial Park
- Shanghai Baoshan Industrial Park
- Shanghai International Auto Town Spare Parts Industrial park

- Shanghai Songjiang Industrial Zone
- Shanghai Kangqiao Industrial Zone
- Shanghai Comprehensive Industrial Development Zone
- Shanghai Chongming Industrial Park
- Shanghai Chemical Industrial Zone

- Shanghai Qingpu Industrial Zone
- Shanghai Jinshan Industrial Zone
- Shanghai Baoshan City Industrial Park
- Shanghai Zizhu Science Park

Shanghai District-level Development Zones

- Shanghai Damaiwan Industrial Zone
- Shanghai Yuepu Industrial Zone
- Shanghai Liantan Industrial Zone

- Supporting Park for Chaogang Heavy Equipment Industry in Lingang
- Shanghai Seaport Comprehensive Economic Development Zone
- Shanghai Nanhui Industrial Park

- Minhang Hi-tech Park in Caohejing Development Zone
- Shanghai Gucun Industrial Park
- Shanghai Putuo Changzheng Industrial Zone

Shanghai District-level Development Zones

- Shanghai Luodian Industrial Zone

- Shanghai Non-ferrous Hi-tech Industrial Park

- Shanghai Fengjing Industrial Park

Table 4.3   *Concluded*

Shanghai District-level Development Zones		
• Shanghai Textile Industry Zone	• Shanghai Gongkang Urban Industrial Park	• Shanghai Xuhang Industrial Zone
• Sub-zone in Fengxian, Shanghai Chemical Industrial Zone	• Torch High Technology Industry Development Center	• Shanghai Rongbei Industrial Zone
• Shanghai Renben Industrial Park	• Shanghai Songjiang Hi-tech Park	• Shanghai Songjiang Science Park
• Shanghai Jiading Industrial Zone	• Caojing Industrial Zone in Jinshan District, Shanghai	• Shanghai Nanotechnology Promotion Center
• Shanghai Shibei New Industrial Zone	• Shanghai Qiangmin Economic Town	• Shanghai Zhangsong Storage Industrial Zone
• Shanghai Zhongshen Baoshui Cangchu Industrial Park	• Shanghai Minhang Electric Industrial Zone	• Shanghai Shangta Industrial Zone
• Shanghai Zhaoxiang Industrial Park	• Zhonggu Industrial Park	• Shanghai Liantang Industrial Park
• Shanghai Zhujiajiao Industrial Park	• Shanghai Huaxin Industrial Park	• Shanghai Baozhen Industrial Zone

Sources: Website of the Shanghai Development Zone, http://www.sidp.gov.cn/; website of investment in Shanghai, http://www.zhaoshang-sh.com/kfqzs/index_3.asp

At the national level, 13 cities have ports, national-level development zones (or Special Economic Zones), and bonded areas at the same time. Any port with the combination of these three types of zones can be called a "Processing-Trade-Port Zone" (PTP Zone). Cities with PTP Zones are obviously more advantageous in attracting foreign investments than other cities in the same province (assuming that Tianjin is part of Hebei Province and Shanghai is part of Jiangsu Province). However, the same research also pointed out that these PTP Zones constitute the major factors that allow these cities to have greater income differences between urban and rural residents than in other cities in the same province. In addition, when making comparisons among these cities with PTP Zones, those with more foreign investments per capita have greater urban-rural income differences.

A common problem with these zones in the zone-port interplay is the lack of attractiveness at a new port district. On the one hand, many industries (e.g., cargo generators) prefer to be close to a port, but that port must have a good connectivity to the world market by possessing adequate services and routes of shipping lines (Figures 4.10 and 4.11). The shipping lines, on the other hand, have similar expectation, that is, to have adequate cargo at the ports of call. Due to this chicken-and-egg problem, the city governments and their zone managements find it difficult to carry out the policy realistically to host the ideal tenants only to their zones, as it may take decades to see the mutual built-up of cargo volume and shipping services. The short-sign governments tend to surrender themselves by loosening their criteria to allow several firms that lack requirements for ports to enter the zones to help their own agenda for political reasons. For example, in Ningbo Port and Shekou Port of Shenzhen, several medium-sized manufacturers of food products not only landed in the development zones there, but also took a piece of coast to build their own terminal with low utility for decades. The long lease issued on lands (compared with the usual 30 years) gave them very lucrative returns when they are eventually "invited" to get compensations and leave after a decade or two of operation, given the high demand of land for port-related activities when the port grows well.

4.6 Port-Based New Cities and their Scales

Thus far, we have discussed issues about ports' spatial leap-forward, the impact of other administrative regions, the assembly of special industries near container ports, the consociation of ports and bonded zones, and the zone-port integration. Given that this work studies the evolution of port-city relationships from the perspective of geographical organization instead of analyzing the ports' economic benefits from the perspective of marine economics, it can thus be inferred that the spatial transformations of major ports in China are actually formulating a series of port-based new cities. This inference is based on two reasons.

First, the spatial leap-forward of many ports does not merely result in new development zones or port districts, but also brings new port-based urban centers to the host cities. This argument can be corroborated by a research on the city of Ningbo conducted by Li et al. (2008). This research argued that multi-temporal land use and urban growth information of three urban districts in Ningbo (Sanjiang, Zhenhai, and Beilun) is extracted from data provided by multi-temporal Landsat MSS, TM, and ETM satellite images. Then, the spatio-temporal characteristics of urban land use growth and the morphological evolutions of these districts are analyzed and the following results are obtained:

1. The speed and intensity of urban land use expansion in Ningbo have continued to increase since 1979, although there are great differences among these three districts.
2. The fluctuation of the fractal dimensions of the urbanized land in each district shows an evident correlation with the spatial expansion of urban land use.
3. The peaks of the intensity index of urbanized land expansion in these three districts tend to move away from the original city center; however, changes in the peak value are different in each district.

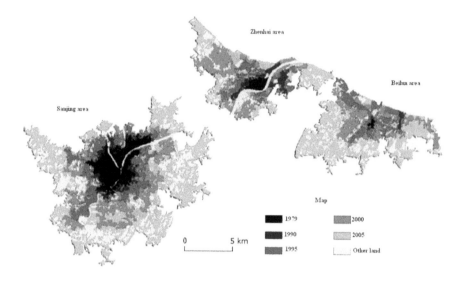

Figure 4.13   Spatial Evolution of Land Use in Different Development Periods of the Port-City Ningbo

Source: Adapted from Figure 1 of Li Jialin et al. (2008)

4. The morphological evolutions of the urbanized land in the three districts are closely related to the spatial shift of Ningbo Port and the development of the port economy.
5. As Ningbo Port gradually changed from a river port to an estuary port and a sea port, Ningbo's morphology experienced several stages of change: from the original river port city with relatively agglomerated development to an estuary port with a city center and a port (Zhenhai Port), a city with two sea ports (Zhenhai and Beilun), and a city with more than two ports. In the future, Ningbo is expected to finally transform into a T-shaped international port metropolis (Li, Zhu, and Zhang 2008).

That study clearly demonstrates that both Zhenhai, which mainly handles bulk cargos, and Beilun, which focuses on containers, development zones and bonded zones, reflect urban expansion and the establishment of new urban centers in Ningbo (Figure 4.13).

The second reason lies in the scales of the port cities in China. Table 4.4 presents information about 10 Chinese port cities with leaped-forward ports as well as several foreign port cities. When comparing major Chinese port cities with foreign ones, we can easily find that either the population or the land area of Chinese large port cities is more similar with that of states or prefectures in foreign countries. Meanwhile, the districts where the ports are located in China are almost as large as the total port cities in foreign countries. Numerous studies have pointed out that the English word "city" should not be a counterpart of "Chengshi" (which means "city") in Chinese, because many Chinese "Chengshi" have already been as large as metropolises. According to the table, six Chinese port cities with new port districts currently exist; their populations or land areas approach or exceed one million or 1000 kilometers. Thus, these cities are truly metropolises rather than cities according to either Chinese or international standards.

Table 4.4   Comparison of the Scales of Port Cities

Chinese port cities	Area (square kilometers)	Population	Location of the new port area	Major functions of the new port	Area (square kilometers)	Population
Shanghai	6,341	13,680,800	Pudong New Area	Container	570	1,875,622
Tianjin	11,305	9,522,793	Binhai New Zone	Container and bulk goods	2,270	1,123,900
Ningbo	9,365	5,604,494	Beilun + Zhenhai	Container	952	580,978
Guangzhou	7,434	7,607,220	Fanyu	Containers and automobiles	786	947,607
Qingdao	10,654	7,493,812	Huangdao	Container and bulk goods	274	316,606
Dalian	12,574	5,720,810	Jinzhou	Container	1,353	702,294
Yingkou	4,970	2,310,783	Bayuquan	Container and bulk goods	268	320,235
Fuzhou	11,968	6,227,327	Fuqing	Container	1,518	1,223,290
Xiamen	1,569	557,431	Haicang	Container and bulk goods	174	28,226
Hong Kong + Shenzhen + Dongguan	1,012 + 1,952 + 2,465 = 5,429	6,880,000 + 8,464,311 + 8,691,000 = 24,035,311	Shenzhen	Container	1,952	8,464,311

Table 4.4   *Concluded*

Host areas of world port cities	Area (square kilometers)	Population	World port cities	Area (square kilometers)	Population
Randstad	Lack of statistical data	6,700,000	Rotterdam	Container and bulk goods	
Johor of Malaysia + Singapore	19,984 + 707 = 20,684	3,300,000 + 4,839,400	Johor Bahru	Container	185	876,000
New York city + New Jersey	930 + 20,295 = 21,215	8,143,197	New Jersey	Container and bulk goods	20,295	8,410,000

Source: Data on population and areas collected from the respective websites of the municipal governments of the cities

4.7   Conclusion

This chapter focused on the processes, characteristics, and impacts of the evolutions of the spatial relationships of port-cities in China. Following the models of port expansions and shifts, the impacts of administrative spatial rescaling on the ports' spatial expansions, the interactions between the ports and the spatial distribution of port-related industries, as well as international space for trade, such as bonded zones, are further discussed. Moreover, the evolutions of Chinese port cities are not only related to trade globalization but also to the dominance of the global supply chains, where "distribution determines production". What differentiates this set-up from traditional producer-driven production is that buyers or their supply chain managers start to consider the place of production, transportation, and the total logistical cost (including the choice of ports for import and export) before products are manufactured. In this model, overseas market and foreland hold the initiative while the "hinterlands" (i.e., areas where products are produced) are being selected. A further point that needs to be emphasized is that PTP Zones have been formulated in several large-scale Chinese port cities. Some PTP Zones are large enough to be called "port-based new cities", and are based on global supply chains. One of the major differences between these cities and their foreign counterparts lies in the management and governance system, which will be discussed in Chapter 6.

Chapter 5
External Relationships of Chinese Port Cities

5.1 Introduction

Specific ports in a country serve as a pivot or a gateway for international trade and communications. In examining these pivots, we can emphasize on the contributions of the host cities to the ports, such as their economic and trade growth potential. The emergence and the increase in the number of ports worldwide are inevitably related to the growing economic and trade systems in which ports serve as connecting points of land and of water transportation corridors. Robinson (1976) and Hilling (1996) suggested that ports should be analyzed as a whole functioning system rather than as sole entities. A port system should consist of the following levels: (I) an intra-port system, (II) a port-hinterland system, (III) a port hinterland-foreland system, (IV) a regional port system, and (V) a total port system.

Many studies had investigated the intra-port system (I), such as analyses on the intra-port system of a specific port and comparisons between different intra-port systems. Research on the regional port system (IV) has also increased, with main focuses on regional inter-port competition and on the evolving distribution of cargo flows. Notteboom (1997), for example, studied the concentration of European container ports in the 1980s and in the 1990s. Charlier (1996) conducted a similar

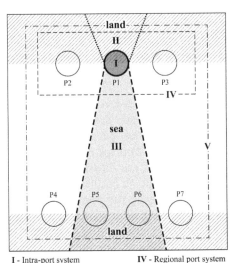

Figure 5.1   Multi-Level Port System

Source: Figure 5.1 from Robinson, R. 1976. Modelling the Port as an Operational System: A Perspective for Research, Economic Geography, 52(1): 718–6. This material is reproduced with permission of John Wiley & Sons, Inc.

I - Intra-port system
II - Port-hinterland system
III - Port hinterland-foreland system
IV - Regional port system
V - Total port system

research on coastal ports in Belgium and Netherlands. Meanwhile, Fremont (2007) found that the distribution of cargo flows from south to north in the European continent is influenced by two factors, namely, the emergence of large ships and the hub-and-spoke approach of large shipping companies. Rodrigue et al. (2007) also analyzed similar situations in North America, while Wang and Jin (2006) focused on Chinese ports.

Several investigations at distinct angles had also been conducted on the port-hinterland system. Well-known cases and models on the relationships between ports and the development in developing countries, such as the model developed by Taaffe et al. (1963) and its later variations such as Hoyle (1968), had focused on the process and on the stages of formation of inland networks, consequently reflecting economic development stages (Rostow 1960) and the contributions of ports and of transportation networks to developing countries. Some geographers and historians, such as those from India and China, had recognized the significant importance of port-hinterland relations and had traced relationship trends in the past few hundred years in their countries (Wu 2006; Banga 1992). These studies revealed the tremendous influences of international trade on economic, cultural, and social development, as well as the effect of geographical variations in a country.

At present, scholars studying the status quo and the development trend of port-hinterland relationships have shifted their focus to the following aspects:

1. Characteristics of various means of transportation and their potential as dedicated logistics corridors (Notteboom 2007; Rodrigue 2007)
2. Integration of land or river transport and seaport businesses and their influences (de Langen 2008)
3. Relationships between inter-port and inter-hinterland competition (Charlier 1996)
4. Effects of hub ports and of the implementation of the hub-and-spoke operation model on the hinterland (Notteboom and Rodrigue 2007; Shi and Zhou 2005; Wang and Jin 2006)

Among the previously mentioned three systems, the high-level port hinterland-foreland system (III), also called the foreland-shipping port system, has seldom been investigated. A number of reasons explain this bias, but shipping system nevertheless deserves considerable research attention. Perhaps by the influence of funding sources, a bulk of early research on ports has focused on shipping economics and on the contribution of ports to the shipping business. As a result, the role of ports is better understood from the shipping perspective than from the city or the national perspective. On this regard, we have seen a recent trend to investigate the port-city competitiveness both in Europe and in Asia, such as a series of studies by the OECD Port-Cities Programme (Merk and Bagis 2013, Merk et al. 2011; 2012; Merk and Hesse 2012,. In the national or the regional level, ports are usually studied in connection with their competition in the hinterlands rather than in the forelands. Belgian scholars, for example, are more interested in analyzing the competition between Antwerp and Rotterdam in the German hinterland, while Korean scholars assess the competition between Shanghai and Pusan in the Chinese hinterland.

Global supply chains (GSCs), especially those related to containerized cargos, have increased and are driven by buyers (Gereffi et al. 2005). When these GSCs obtain products or search for appropriate producers in developing countries, their targets are not always product sources or manufacturers but rather suitable countries or regions for production. Thus, some of them establish their own factories before the products are manufactured. Coastal port cities in developing countries with good connectivity with the rest of the world consequently become favorable sites for GSCs to cluster their export-oriented manufacturers. In the foreland perspective, the emergence of buyer-driven GSCs forwards the question: "Which port cities or areas are qualified to be chosen by GSCs?" The answer to this question has multi-faceted considerations. The connectivity of shipping lines in a port city or in an area is a definite factor. Thus, foreland-shipping-port systems should be examined. Findings of this study will be essential to port city governments in China, as ports serve as their leverage in the competition for higher trade volume and thus higher GDP. The succeeding section analyzes connectivity differences among Chinese ports with international shipping lines and the reasons behind the said differences.

Since 1978, China has undergone significant transformation from an economically closed country to one that relies on foreign trade. Over 80 percent of the foreign trade in China is accomplished through its mainland ports. In 2007, China had 38 ports that regularly cater to international container shipping lines (excluding those in Hong Kong and their annual throughput amounted to 80 million TEU. Included in the huge amount of cargo flows were products sourced by final sales market operators (e.g. Wal-Mart and Carrefour), produced and distributed by transnational corporations (e.g. Nike and Samsung), and delivered by small and medium-sized enterprises. GSC managers (e.g. Li and Fung) who serve as agents for some multi-national corporations (e.g. Coca-Cola and Toys "R" Us) seek suitable manufacturers for buyer-chosen products and identify areas in China and in Asia for the assembly and the delivery of specific product components. Therefore, various ports not only manage cargo flows of final products between China and foreland markets, but also those of semi-final products among countries involved in component production.

Although GSCs have different organizational patterns (Gereffi et al. 2005), they similarly require production sites, particularly in China, to have geographical proximity to ports. This requirement is the reason why many large multi-national corporations concentrate in a few portside national-level development zones. These zones are mainly located in coastal cities in eastern China where more than 92 percent of the total international trade (in terms of value) is accomplished by several companies. The high-level clustering of enterprises, particularly of export-oriented manufacturing firms, is relevant to the international connection among ports. In contrast to large coastal iron and steel or petrochemical enterprises that require dedicated wharfs, export-oriented manufacturing enterprises and container terminals are both *independent* and *interdependent*. These enterprises are independent because each firm that generates cargo flows does not have a substantial relationship, such as share-holding, with the port, but only generates demand for the port. By contrast, these enterprises are interdependent because, on one hand, a fledgling container port may only have a small cluster of enterprises or create business cooperation with only a few shipping companies and terminal operators. The establishment of new shipping lines may attract more enterprises. On the other hand, if enterprises fail to cluster in a port, shipping services may be reduced or even abolished. If the number of shipping line companies that launch new services in a port increase, the scale of enterprise clustering in that port becomes sufficiently large to generate profits for both the port and the shipping lines. Furthermore, if a shipping line company invests in a port, the company regards the port as a permanent growth pole of business where enterprises cluster in and around the city.

However, significant differences are observed in terms of scale, stage of development, and number of external shipping lines among the 38 coastal ports in China with international shipping connections. Any enterprise willing to establish a foothold and build a production base in coastal cities in China must consider adopting the foreland-shipping-port system to find a suitable place for future development. Under this system, enterprises not only expect to know (1) the availability of shipping lines connected with their foreland markets and (2) the differences in shipping fees, but also (3) the frequency of shipping services and (4) the number of service providers. These four conditions determine if GSC is reliable and has available alternatives in terms of networks and markets that can be substituted. The first condition determines the connection of a port to specific markets, while the third and the fourth conditions constitute the connectivity of a port with other ports. The section below compares the connectivity of different major Chinese port cities with international shipping lines to major foreland markets worldwide.

5.2   External Connectivity of Chinese Ports

5.2.1   Research method

The connectivity of a port is not one-dimensional, but two-faceted:

1. Market coverage

Considering significant variations among Chinese ports in terms of foreland coverage, the market should be divided into regions (rather than cities or countries) to ensure consistency in the analysis. Thus, based on the foreign trade in

China, this study divides the world into the following eight foreland submarkets: (1) Far East – North America (FE–NA), (2) Far East – Europe (FE–EU), (3) Far East – Australasia (FE–AU), (4) Far East – Middle East (FE–ME), (5) Far East – Africa (FE–AF), (6) Far East – South America (FE–SA), (7) Intra-Asia, and (8) round-the-world lines and others. The connectivity of a port to these areas is determined by the availability and by the number of shipping lines. To show the differences in the concentration of market coverage of the ports in China, we use the Herfindahl-Hirschman Index (HHI), a comprehensive indicator of market concentration. HHI is the sum of the squares of market shares of the total number of enterprises in a certain market j.

$$HHI = \sum_{i=1}^{N} (X_i / X)^2 = \sum_{i=1}^{N} S_i^2 \quad , \quad (1)$$

where

X: total size of a certain market (i.e., total number of shipping services)
X_i: total share of port i in the market (i.e., number of shipping services)
$S_i = X_i$, where X is the market share of port i
N: the number of ports

The above equation yields the market share of port i in each submarket j. The sum of all HHI_{ij} is then divided by the number of submarkets. Finally, the overall indicator of the foreland market coverage of every port M_j is described as

$$M_i = \frac{1}{m} \sum_{j=1}^{m} (w_j HHI_{ij}) \quad , \quad (2)$$

where m represents the number of submarkets and w_j is the weight of each submarket (i.e., 1 in this study).

Similar with the standardized HHI, the maximum value of the market coverage index is 1 or 100 percent and the minimum value is 0. This index is affected by the number of participants or ports in this study.

2. Frequency of service

For clients, once shipping lines cover a market, the frequency of service becomes a key consideration because it can directly affect connectedness and efficiency of the entire logistics chain. In this research, the total number of shipping services targeted by each port for each of its submarket in a month is regarded as a statistical unit. A shipping line may provide several shipping services per week, a service every two weeks, or a service per month. Given that different shipping companies can use the same container ship in a specific scheduled service (usually also belonging to the same strategic alliance) but report them separately, efforts have been devoted in identifying the deployment of container ships and of similar port rotation sequences. The frequency of services for specified markets must be equivalent to the total number of services necessary at the studied ports.

5.2.2   Data source

Our data are mostly obtained from the Chinese and from the Hong Kong versions of the *Shipping Gazette*, which provides information for clients on shipping line services in China, particularly scheduled shipping of each shipping line company, sailing schedule, and the voyage distance of each liner that arrives at or departs from each Chinese port. Considering the difficulty in collecting and in processing shipping service information for an entire year and the possible omission of some biweekly services, we use month as our time unit in calculating the frequency of service. To avoid busy seasons such as September and October when manufacturers ship products for Christmas in Europe and in the US and unstable off-seasons such as January and February when Chinese people celebrate the Spring Festival, we select the month of July for our analysis. To determine the annual trend, we collect data on the same month in two consecutive years (2006 and 2007).

Given that the shipping schedule and the number of services provided by various shipping companies do not report on instances of sharing ships with other companies, but instead label ships with different codes (i.e. code sharing), we spend an enormous amount of time identifying all code-sharing cases and eliminating repetition by checking vessel names and shipping schedules. In addition, to reduce and to avoid the incorrect inclusion of scheduled shipping services that were not realized or were later changed, we utilize real-time information of major ports to check the arrival and the departure of ships.

With the connection of Hong Kong to mainland China and its integration into the port group system in the Pearl River Delta we include Hong Kong in our analysis. Thus, our data comprise 25 Chinese ports with outbound shipping services and 63 shipping line companies that provide international shipping services (Figure 5.2).

5.2.3   Findings

Table 5.1 shows the number of services covered by 25 Chinese ports from the Dandong Port of Liaoning province in Northeast China to the Fangchenggang Port of Guangxi province in Southwest China. The services are sorted according to submarket. Several scenarios are identified:

1.   Only 12 ports served the intra-Asia market and had no direct connection with other submarkets.

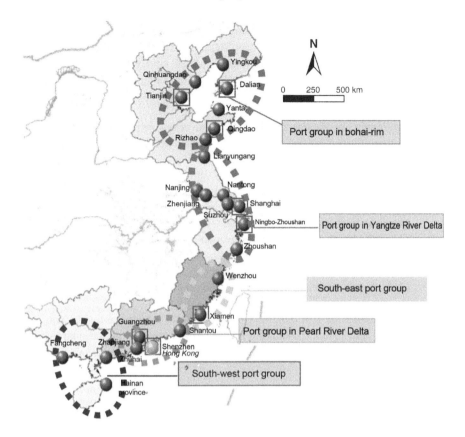

Figure 5.2   Distribution of Major Coastal Ports in China

2. The frequency of services provided to the intra-Asia region in the Yangtze and in the Pearl River Deltas validated the prominent position of these two deltas. The ports of Shanghai and of Ningbo served as the hub of the Yangtze River Delta, while the ports of Hong Kong and Shenzhen served as the hub of the Pearl River Delta.

3. The three major ports in North China, namely, Dalian, Tianjin, and Qingdao did not have as many services as Shanghai and Hong Kong in the intra-Asia market, but had more intra-Asian connections than Ningbo and Shenzhen.

4. Surprisingly, Xiamen Port covered all submarkets except Africa, providing more shipping services to Europe than to any other port in North China, including Qingdao, Tianjin, and Dalian. The number of services provided to North America was second only to that from Qingdao, which had already become the regional hub of Maersk-AP Moller. Xiamen Port ranked sixth in China in terms of the total number of services to North America.

Another surprising finding was the Guangzhou port that had a total of 68 shipping services, covering all submarkets except South America. Overall, the Guangzhou and the Xiamen ports demonstrated higher market coverage than that of other ports in North China, but lower than that of Hong Kong, Shanghai, Shenzhen, and Ningbo. Despite the high market coverage of these ports in some areas, they indicated low frequency of service.

The following phenomena are observed if the coverage is calculated according to the number of services (and not the scale) of each port (Figure 5.3):

1. Shanghai and Hong Kong had highly similar market coverage, as well as their competitors from another region, Shenzhen and Ningbo.
2. Several ports exhibited interesting market coverage characteristics. For example, Fuzhou directly connected with Asia, Africa, and the Middle East, although the coverage of the latter two submarkets was small. A similar situation was observed in Lianyungang (Asia, North America, the Middle East, and others), Yantai (North America and intra-Asia), and in Dalian (Asia, North America, Middle East, and others).

Figure 5.3 presents the combination of the frequency of service results with M_i in formula (2), displaying the ranking and the differences in international market connectivity (July 2007) of all Chinese ports with international container shipping lines.

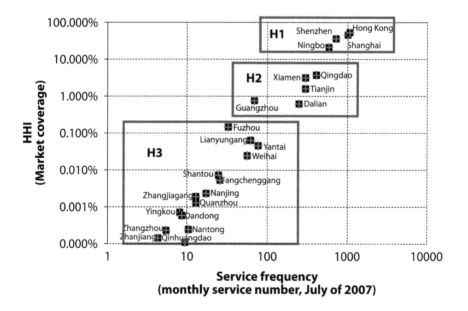

Figure 5.3   **International Connectivity and Coverage of Chinese Coastal Container Ports with Overseas Shipping Services**

Table 5.1 Number of International Liner Services Provided at Coastal Ports in China in July 2007 (According to Different Overseas Submarkets)

Overseas Submarkets	Europe	North America	Middle East	Middle and South America	Africa	Australasia	Asia (except Middle East)	Other	Total service number
Number of liner companies	24	23	34	16	12	24	46	26	63
Number of Chinese docking ports	9	11	11	7	8	7	24	10	24
Monthly number of services	210.0	261.4	141.4	68.6	42.1	52.4	1,258.3	88.7	2,123.0
Dandong							8.6		8.6
Yingkou							8.6		8.6
Dalian	21.4	21.4	12.9				186.4	4.3	246.4
Qinhuangdao							4.3		4.3
Tianjin	42.9	25.7	25.7	4.3	4.3		186.7	8.6	298.1
Yantai		4.3					72.9		77.1
Weihai							55.7		55.7
Qingdao	51.4	51.4	34.3	4.3	4.3	15.0	244.3	4.3	409.3
Lianyungang		4.3	4.3				49.3	4.3	62.1
Nanjing							17.1		17.1
Nantong							4.3		4.3
Zhangjiagang							17.1		17.1
Shanghai	141.4	141.4	90.0	60.0	30.0	39.6	437.1	67.3	1,006.9
Ningbo	128.6	98.6	77.1	34.3	21.4	17.1	162.9	40.7	580.7
Fuzhou			4.3		4.3		23.8		32.4
Quanzhou							12.9		12.9

Table 5.1   *Concluded*

Overseas Submarkets	Europe	North America	Middle East	Middle and South America	Africa	Australasia	Asia (except Middle East)	Other	Total service number
Xiamen	72.9	38.6	4.3	4.3		12.9	115.7	17.1	265.7
Zhangzhou							4.3		4.3
Shantou							17.1		17.1
Shenzhen	184.3	158.6	81.4	42.9	25.7	17.1	137.1	64.3	711.4
Hong Kong	162.9	150.0	107.1	51.4	30.0	46.0	439.3	77.1	1,063.9
Guangzhou	17.1	8.6	4.3		8.6	4.3	21.4	4.3	68.6
Zhanjiang							4.3		4.3
Fangchenggang							25.7		25.7
Share of submarket	9.9%	12.3%	6.7%	3.2%	2.0%	2.5%	59.3%	4.2%	100.0%

Source: Chinese and Hong Kong editions of the Shipping Gazette

In this figure, we classify all ports into three hypothetical hierarchies. The top hierarchy consists of Hong Kong, Shanghai, Shenzhen, and Ningbo that demonstrate the strongest connectivity. The bottom hierarchy, with the weakest international connectivity, consists of 15 ports that have direct connections to (non-Chinese) Asian ports only, suggesting their regional and feeder status and trade characteristics (Figure 5.2). In general, the geographic coverage of this hierarchy is highly concentrated in a single or a few region(s) (mainly intra-Asian) that reflect major differences in the frequency of services. Interestingly, however, Dalian, Guangzhou, Qingdao, Tianjin, and Xiamen were classified into a distinguished group with visible characteristics overtly different from those of the top and bottom hierarchies. The cities in this group are regarded as second- and third-tier Chinese cities with relatively strong connectivity to major world markets and high potential for further development. Each port has its own merit in terms of connecting with selective markets partly because of their geographical location along the Chinese coast. For example, Dalian, Qingdao, and Tianjin connected much better with ports in Japan and in Korea, while Xiamen with Southeast Asian ports. Although geographical proximity is not a surprising factor if analyzed separately, the absence of differentiation of connectivity in the hierarchy of highest connectivity validates the global coverage of shipping connectivity in two most important mega-city regions in China, namely, the Pearl and the Yangtze River Deltas by two pairs of ports, Hong Kong-Shenzhen and Shanghai-Ningbo, respectively.

To increase accuracy, we calculate the average daily number of services of all Chinese shipping companies serving overseas markets and compare the data gathered in July 2006 and those in July 2007. The following main points are worth noting:

1. The overall patterns of shipping services in China for two years are consistent, indicating the stability of overall connectivity.
2. Guangzhou increased the most on the number of services provided because of the establishment of the new Nansha International Deepwater Terminal. However, the total number of services of Guangzhou was insufficient to change the overall pattern of shipping services. Rizhao observed the most significant decrease due to the cancellation of a shipping line to the intra-Asia market in 2006.
3. Shanghai and Ningbo witnessed the same increase in the number of services, as well as Dalian and Tianjin, suggesting that consistency in regional service adjustment translates to consistent changes in the connectivity of port cities.
4. The two major ports of the Pearl River Delta, namely, Hong Kong and Shenzhen, remained the same, similarly suggesting that changes in the amount of service of the ports in the same region are consistent.

According to a comparison between 2006 and 2007 data and to the results of our field survey, ports with weak connectivity were sensitive to requests for new

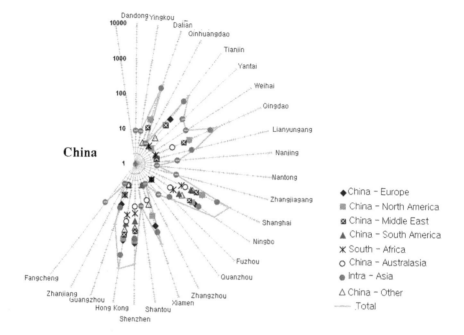

**Figure 5.4   Frequency Distribution of Connectivity between International
Container Routes and Chinese Ports (Number of Services in
July 2007)**

vessels from liner companies. Conversely, shipping line companies were cautious
in establishing international shipping lines in small and medium-sized ports. Thus,
new lines are not easily developed in these ports. However, in major ports in a
region, shipping companies commonly adjust their shipping lines and their ship
types or associate themselves with other liners to share shipping space rather than
changing the connectivity of the ports by adding more shipping lines. A comparison
on line sharing between several shipping line companies in July 2006 and in July
2007 revealed that many shipping line companies in each submarket adjusted
their shipping lines (e.g., changed the number of lines or their destinations). These
adjustments were substantial for individual companies. However, the changes
were insufficient to change the overall overseas market connectivity.

5.2.4   Significance of external connectivity of port cities

Figure 5.4 outlines the general relationship between port cities with outbound
shipping lines in China and the number of lines to all submarkets worldwide in
2007. For individual container shipping line companies, the relationship illustrated
in the figure is not significantly useful because these companies are more concerned

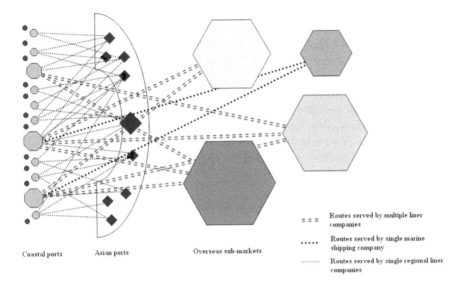

Coastal ports   Asian ports   Overseas sub-markets

= = = Routes served by multiple liner companies

..... Routes served by single marine shipping company

.... Routes served by single regional liner companies

Figure 5.5   Conceptual Model of the Multi-level Overseas Connection System of Chinese Ports

on details such as surplus capability of each line to make appropriate arrangements in every port. However, for cities and provinces with ports, this figure reflects the following important characteristics: (1) the significant differences among channels of foreign trade in different regions in China and (2) the roles of individual ports in the national port hierarchical system.

On one hand, Figure 5.4 indicates that foreign trade channels generally reflect foreign trade partners/markets in different regions. On the other hand, the operations of shipping lines are critical for foreign trade in each region because any indirect connection, land transport, or sea transshipment quickly increases the overall cost of the supply chain. This situation is very serious in China because hidden costs vary from one customs department to another and cross-province shipments by land are highly uncertain. Buyer-oriented purchasers also prefer seeking manufacturing factories in port districts with direct and frequent shipping services those in places with inconvenient transportation. Therefore, regular, direct, and frequent services results in further clustering of suppliers near the port districts.

Regarding the port hierarchy in China, after several decades of economic development after the reform and the opening-up, Chinese ports have basically been shaped into a spatial hierarchy. Figure 5.5 shows the characteristics of this hierarchy system. The ports are divided into four types according to their connectedness with overseas markets, namely, (1) local ports without international shipping lines, (2) regional ports with lines connecting with the intra-Asian market by a single shipping company, (3) international artery ports with connections to specific overseas markets, and (4) international hub ports with connections to all major international markets.

Based on our 2007 data, the validity of the aforementioned two characteristics of Chinese ports with external links cannot be proven. However, the fact that Chinese ports in different regions were affected by the 2008 global financial tsunami at different degrees reflects the varied levels of external connectivity of Chinese ports. For example, under the same period, Shenzhen and Hong Kong were most affected by the slow progress of the North American market (container shipments decreased by 16 percent to 19 percent in 2009), whereas Tianjin and Guangzhou were least affected (container shipments increased due to growth in the domestic market).

5.3   Port-Hinterland Relationships in Fujian Province

The port-hinterland relationship shown in Figure 5.1 is re-examined. Traditional economic geography and transportation studies define "hinterland" as the area where the flow of goods into a certain port or airport originates. Thus, hinterland is the source of cargos received by ports. The larger the hinterland, the stronger is its economy. Besides, the more resources the hinterland has, the greater development potential is exhibited by the port. Given that a region or country may have more than one port, the hinterland is classified as either direct (i.e. a resource area dominated by a single port) or indirect (i.e. a source of cargos shared by many competing ports) hinterland. According to the previously mentioned concept of the hinterland and to the division of different hinterlands, every geographical analysis of ports unavoidably demonstrates the potential of the hinterland to predict future development. The obtained data are often regarded as a basic reference in formulating a port development strategy.

However, the above methodology has many limitations. First, some ports are hinterland-based, whereas some are transshipment-based. Some ports do not even rely on hinterlands such as the Kaohsiung Port in Taiwan. Second, although most hinterland-based ports are based on their direct hinterland, that is, the economy of the city or the region where the port is located, some ports rely on indirect hinterlands rather than on the direct one, such as the Qinhuangdao Port because the city of Qinhuangdao and its surrounding areas lack a significant amount of goods. Third, the most important point is that the hinterland is merely a part of the supply chain from the place of production to the market. Under economic globalization, the relationship between the market and the source of raw materials and the processing location is changing fast. The conventional one-way description of "hinterland to market" is no longer reconsidered and redefined. The "hinterland" in the traditional one-way flow of trade may become the "market" in a two-way flow of trade. In this sense, the hinterland is not necessarily a place that connects with a port by land. For example, the hinterland of Fangchenggang Port, which is adjacent to the source of ore in Australia, relies on its country and on its market in Southwest China. Similarly, if a port in Fujian imports local specialties from Taiwan and then transports them by train to markets in the mainland, Taiwan becomes the hinterland of the port and the provinces and cities in mainland China become its market. The transition of China from the current export-oriented economic structure to one with balanced imports and exports facilitates the emergence of this two-way relationship between the hinterland and the market.

Moreover, components of a product are commonly outsourced and produced in various places in different countries to take advantage of low cost but high-quality labor and/or materials. These components are often assembled into final products in one location near a good regional hub port before being shipped to global market. From the GSC perspective, this process is a multi-hinterland operation providing hub ports with a certain merit as they connect with multiple production cities in a region with multiple consumption cities globally (Wang and Cheng 2010; 2014). As a result of these practices, the foreland connectivity analysis presented earlier in this chapter becomes part of "hinterland analysis" if the data are about trade rather than about vessel schedules.

If a port or a group of ports in a region has diverse and multi-directional trade relationships, defining and analyzing simple one-way hinterland-market systems, as shown in Figure 5.1, will be misleading. If several ports exist in a region and each of them handles specific cargos (e.g. bulk cargos such as petroleum products and ore and container cargos usually loaded or unloaded in different ports or port districts), the hinterland of each port should be analyzed and discussed according to cargo type and to cargo flow directions. Therefore, simply discussing the extent of the hinterland of a port is meaningless. For example, a broad delimitation of the hinterland of Tianjin and Dalian Ports will be slightly logical.

5.3.1 Analysis of the hinterland of Xiamen Port

In this section, we use the dynamic relationships between Xiamen Port in Fujian and its market and hinterland to identify changing hinterland-market relationships. Given the unavailability of data, analysis of the Xiamen Port only covers from 2000 to 2005. First, we examine container shipments. Xiamen is the leading port for the international trade of container cargos in Fujian. In 2005, international container cargos handled in the Xiamen Port accounted to 83 percent of the total throughput in the province. To compare the market and the hinterland connected by water for container cargos in the Xiamen Port in 2000 and in 2005, Figure 5.6 and 5.7, respectively, show the entry and the exit of empty containers via the Xiamen Port. Percentages are used to show the magnitude of the entry and the exit of the empty containers.

Figure 5.6 and 5.7 reflect three important characteristics during the five-year period.

1. For the entry of empty containers, in 2000, 63 percent of empty containers were from offshore international markets and 33 percent or nearly one third of them were from Kaohsiung in Taiwan. Supposing that fully loaded containers were returned the same way and in the same percentage, which is generally true in China because the customs department has a tight control

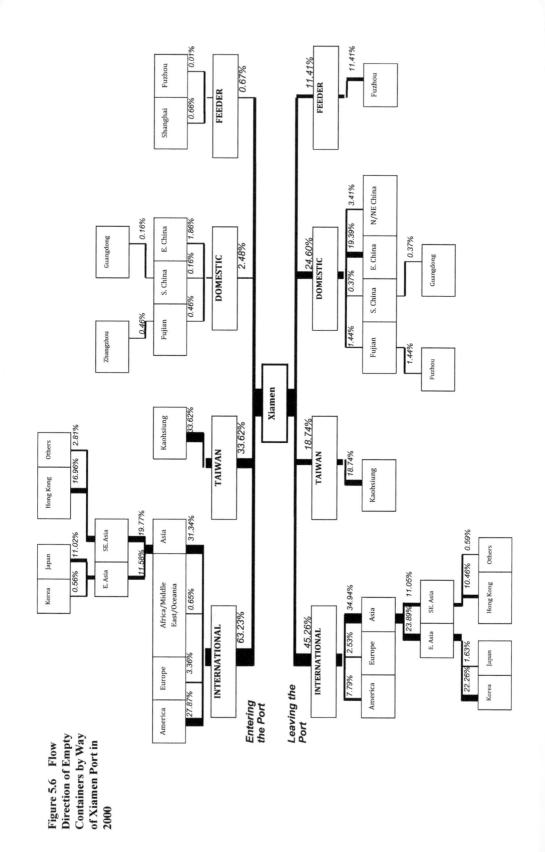

Figure 5.6 Flow Direction of Empty Containers by Way of Xiamen Port in 2000

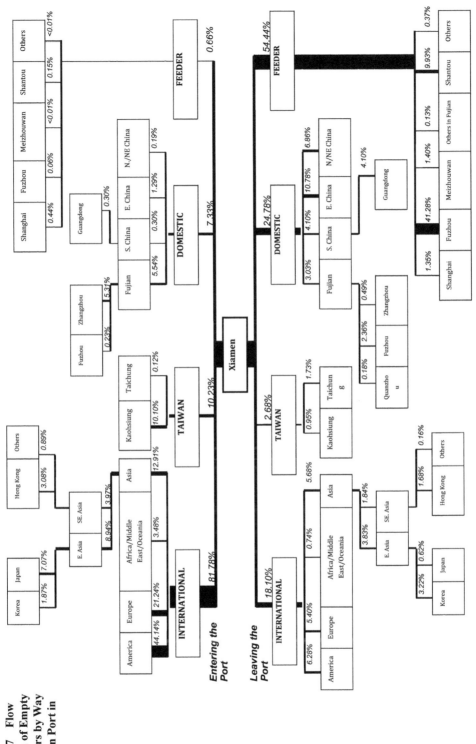

Figure 5.7 Flow Direction of Empty Containers by Way of Xiamen Port in 2005

on these containers as "transport equipment", Kaohsiung Port proves to be the largest transshipment hub for international import and export of goods at the Xiamen Port. This situation significantly changed in 2005 when only 10 percent of the containers entering Xiamen Port were from Kaohsiung and the remaining 80 percent were directly from offshore international markets.

2. For the exit of empty containers, in 2000, only 11 percent of internationally traded empty containers reached another Chinese port (Fuzhou Port) by inland feeder. In 2005, this percentage increased to 54 percent, along with the number of ports connected by inland feeders.

3. During the same period, the percentage of outward empty containers for domestic consumption from the Xiamen Port remained almost the same (approximately 25 percent).

The above-listed characteristics depict an important process, that is, the transformation of the Xiamen Port from a regional artery port to a regional hub port. In other words, Xiamen Port, the largest container port on the west coast of the Taiwan Straits, evolves from a port that relies on cargos from its direct hinterland to one that functions as a regional hub. The hinterland-market interaction in the Xiamen Port is mainly reflected in the exchange of goods between coastal areas and offshore markets, rather than between the port and the landside hinterland. The direct hinterlands of this region include Xiamen, Quanzhou, and Zhangzhou, whereas Shantou and Fuzhou are its indirect hinterland.

5.3.2   Land-connected hinterland of Fujian ports on west coast of Taiwan Straits

Given that we lack precise data on cargo flows through railways and highways, we cannot conduct the same analysis for these flows as that above. However, the data collected indicate that the range of port hinterlands on the west side of the Taiwan Straits is evident and has slightly changed. First, the hinterland of Xiamen Port includes Xiamen, Quanzhou, Zhnagzhou, and Longyan from the traditional point of view (ranked by the total amount of cargos). The indirect hinterland includes East Guangdong (mainly Shangtou), Jiangxi, and other cities and areas in Fujian. The hinterland in Quanzhou Port is very small and basically includes the city proper, Sanming, and Putian. The hinterland of Fuzhou Port mainly includes Fuzhou proper, Nanping, Sanming, Ningde, and Putian. Apart from the Fujian province, the hinterland serving the port group on the west side of the Taiwan Straits includes East Guanggong, Jiangxi, and the southernmost part of Zhejiang. Among these three areas, Shantou and the southernmost part of Zhejiang are mainly connected with ports in Fujian via coastal shipping and feeders of international shipping lines. Therefore, the main landside hinterland of ports in the coastal areas of Fujian is located alongside the northward land transportation corridors of the ports.

The *Pan-Pearl River Delta Highway Planning*, the *Plan of a Highway Network in the Economic Zone on the West Side of Taiwan Straits*, and the *Comprehensive Transport Planning of Fujian Province* indicate that highways are less crowded than railways. Considering that the entry and exit of cargos in Chinese ports involve

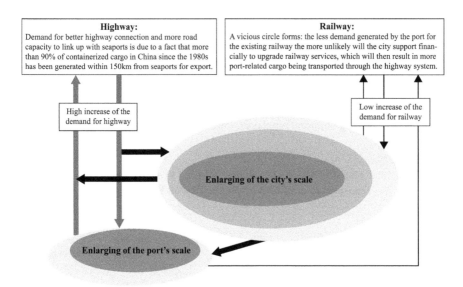

Highway:
Demand for better highway connection and more road capacity to link up with seaports is due to a fact that more than 90% of containerized cargo in China since the 1980s has been generated within 150km from seaports for export.

Railway:
A vicious circle forms: the less demand generated by the port for the existing railway the more unlikely will the city support financially to upgrade railway services, which will then result in more port-related cargo being transported through the highway system.

High increase of the demand for highway

Low increase of the demand for railway

Enlarging of the city's scale

Enlarging of the port's scale

Figure 5.8 **Schematic View of the Unbalanced Development Cycle of Export-Reliant Coastal Cities in China**

passing through highways rather than railways in China, an unbalanced development cycle exists in the process of port-highway-railway development (Figure 5.8). In this process, a positive interaction has been formed among highways, cities, and ports, which tend to grow proportionately. However, the relationships among railways, cities, and ports exhibit the opposite. In the process of port containerization and of rapid seaward urbanization in cities, railways become increasingly less important in transporting cargos. In addition, constructing railways requires long-term investments and a long period of time, and several objective restrictions have been imposed on both private and foreign railway investors. Therefore, Fujian is insufficiently motivated to update and to improve the railway market, although sufficient resources are available to enhance underdeveloped railway system.

From a wider perspective, the unbalanced development cycle described above is only part of the unbalanced development of the overall national transport system. Table 5.2 shows that port, air, pipe, and highway transport have all developed faster than railway transport in China over the past half a century, specifically in the past 30 years. Ports have rapidly developed as a result of globalization and of low water transport fees. However, apart from the flexibility of highways in terms of coverage and the rapid seaward urbanization in China, the similarly rapid development of highways was due to two important reasons. First, in contrast to building railways, highway investment can be segmented. In other words, highways can be built by sections and investments can be made by different investors at different time periods. Therefore, investors in highway construction have instant returns on investment even with the completion of only one section. Highway investors also do not need to invest in vehicles, another important component of the highway system, as vehicle owners buy the vehicles and are thus an integral part of this mode of transportation. By contrast, the railway system only provides services for users who do not have access to roads. Road maintenance fees paid by vehicle owners help promote the further development of roads. Second, for a long time, users of highways have not been held liable on environmental pollution caused by highway transport. The society shoulders this hidden environmental cost, unconsciously benefiting highway users.

Table 5.2   Comparison of the Development Speed of All Means of
             Transportation in China over the Past 60 Years

Main means of transportation	1949	1978	2006	Annual increase rate	Growth rate over past 30 years
Railway:					
Mileage (10,000 km)	2.18	5.17	7.71	1.5%	50%
Passengers (10,000)	10,291	81,491	125,656		
Highway					
Mileage (10,000 km)	8.07	89	345.7	5%	Four times
High-speed highway	0	0	5.36 (2007)		
Passengers (10,000)	1,809	149,229	186,0487		More than 10 times
Port:					
Berths weighing more than 10,000 tons		133	1,203		
Containers (TEU)	1,100 (tons)	<1000	93.61 million	50%	94,000 times
Aviation:					
Number of airports	40 (airstrips)	79 (1980)	147		Two times
Passenger transport	1.04 (1950)	343	18,518.5 (2007)	25%	540 times
Pipe:					
Length (km)	(from 1,958)	8,300	48,200	6.5%	Nearly six times

Source: China Infobank

5.4 Conclusion

This chapter analyzes the connectivity of container ports in China with foreland markets by using two indicators, namely, frequency of service and market coverage. The analysis was conducted from perspective of the shipper and considered the characteristics of the port-foreland market system. Different from previous studies of maritime geography that focus how ports are inserted or linked by shipping routes or network, this analysis attempts to reveal from a port-city point of view, how they are connected to their foreland. Our analysis showed that the shipping connectivity of a port is a stable and reliable indicator that can reflect certain differences among ports in China compared with other measurements. This indicator also served to classify the ports into different types according to their connectivity rather than adopting the approaches of shipping companies on maritime service provision (e.g. hub-and-spoke). A comparison between 2006 and 2007 data revealed that shipping companies considered the overall stability of the number of services and the market coverage of each port when they perform adjustments on their shipping lines and on their cooperative relationships with other shipping companies.

The data in this research were obtained from shipping lines that regard Chinese ports as origin or as destination and do not involve cargo-specific information or services from ports outside China. Similarly, the indirect connectivity by transshipment in hub ports in other countries was excluded from the calculation. This aspect can be further examined and discussed in future research, as many shipments in a hub port may have diverse effects to the city where it is located. On one hand, real connectivity may not be as high as expected if the demand for ships only slightly contributes to the provision of "extra capacity" on ships for locally generated cargo. On the other hand, the extra capacity of transshipment vessels may lead to a concentration of home-based logistics activities in the hub city, resulting in some services to be subjected to last-minute changes in cargo destinations or other demand adjustments.

In addition, the port-foreland connectivity research presented in this chapter can serve as the first part of a time-series research in the future. Given that the relationship between China and international trade is ever-changing, we can use similar data in the conduct of comparative studies tracing the "physical connection" of various regions and cities in China with the rest of the world. The calculation method employed in this research can be applied to analyses on the external connectivity of ports in other regions and countries and on air transport connectivity. Particularly, a city in a developing country with high regional and global connectivity via shipping lines is merited by the consideration of GSCs as an assembly node for multi-hinterland production.

Overall, connectivity is crucial for cities to boost their development in today's "networked world". Thus, connectivity by shipping services is beyond a physical trade link and may lead to some cultural footprints between countries and regions. An example is the concentration of African and Middle Eastern merchants in

Guangzhou, which should be regarded as an important and interesting interplay between strong shipping connectivity and the newly established trade between Guangzhou and the two regions. This high connectivity may be established as well between a city and an entire continent.

In the second section of this chapter, we analyze the dynamic evolution of port hinterlands in China by considering the Fujian province as an example. We observed the three important facts in the land-connected hinterland development in China over the past 30 years. First, the direct hinterland of major container ports is small, such as the hinterland of Tianjin Port. Though the Tianjin Port had been highly stable since 1995, approximately 50 percent of the port's cargos were from Binhai New Area near the port, 45 percent were from other places in Tianjin, Beijing, and in the Hebei province, and 5 percent were from more distant places. Second, railways only slightly affected the transportation of containerized cargo in international trade. As an example, less than 1.5 percent of the containers in large ports, such as Shanghai and Shenzhen, were transported by trains. Third, railways may benefit from the expansion of the Chinese domestic market. However, the accumulation of merits will take a long time. Containerization in China still has a long way to go because railway operators presently prefer transporting human passengers for various historical reasons (Wang et al. 2012). Nevertheless, coastal feeders for domestic trade have quickly developed, largely because after the three-decade establishment of export-oriented manufacturing businesses, many of which are OEM, these makers of consumer goods have begun serving China's increasing consumption rate. Export-oriented manufacturers are highly concentrated along the coast, reflecting the trend of further economic development in the coastal areas in China, as further discussed in Chapter 8.

Chapter 6
Port-City Governance Relations

6.1 Governance Relationships Between the Port and its Host City

The seashore is the physical location and the resource provider for ports. Normally, important sea shorelines are governed and managed by the state government. Different countries have their distinct history, and the roles of their government units vary. Therefore, the management and the operation of ports in each country, in terms of port waters, coastline, land area, and related land and water facilities, among others, differ. The policies of many countries regarding these aspects are changing. Shorelines, generally in most economic systems and particularly in modern societies, are recognized as state-owned resources. However, policies and methods of shoreline management, port planning, construction, and operation differ among countries, as well as the roles played by government units at different levels. Port planning, construction, management, and other activities carried out by government units are collectively called port governance.

Governance goes beyond administration and refers to a diversified management in which corporations, non-government organizations, trade unions, and other communities are involved. Port governance is a kind of collaborative management of a particular transport service sector that includes various ways of policy- and decision-making across sectors and spatial scales. The port sector has become a diversified community (port community) while in the process of modernization. Technical and economic requirements of the transportation and logistics industry, as well as market requirements of trade (particularly international trade), compel ports to carry out structural reforms and adjustments. Structural reforms of ports in many countries have followed two basic trends, namely, privatization and decentralization. Privatization refers to the process in which ports are no longer monopolized by state-owned enterprises, whereas decentralization refers to the dispersion and to the localization of the rights of planning, management, and administration and to the independent operation and self-financing of terminals or wharfs. These two basic trends result in a general tendency, which is the diversification of port governance.

In international academic circles, port governance has been a popular topic for research. The analysis of port governance has two perspectives. One is the economics perspective of port management, which focuses on the influences of different port governance systems on economic efficiency and competitiveness of ports. Studies conducted by Brooks (2004), Notteboom (2005), Pallis and Syriopoulos (2007), Brooks and Cullinane (2007), and others have emphasized on the economics perspective. Property ownership is the first subject to be considered in these investigations. A World Bank research paper (2003) classifies ports worldwide into four types based on ownership (Table 6.1).

Table 6.1   Port Management Structure and Classification Based on Port Ownership

Port Management Models

Types of ports	Infrastructure	Superstructure	Port labor	Other functions
Public service port	Public	Public	Public	Majority Public
Tool port	Public	Public	Private	Public/Private
Landlord port	Public	Private	Private	Public/Private
Private service port	Private	Private	Private	Majority Private

Source: World Bank Reform Toolkit: Module 3 Alternative Port Management Structures and Ownership Models, 2003

The second perspective in the analysis of port governance is more macroscopic. Port governance is regarded as an intermediate level in a multi-dimensional governance structure. Figure 6.1 is a tri-dimensional port governance model that the author proposed in 2004 (Wang, J., Ng, A., and Olivier, D. 2004). The diagram has three dimensions, namely, the vertical axis that is the domain of governance and the two horizontal axes that refer to the complexity of the logistical functions of the port and the diversity of stakeholders involved in port operations. The governance domain indicates that port operation involves a number of levels such as the international community, national, provincial, local, and municipal government units, and port authority, district, and berths. Governance seeks to express power distribution along themes of *spatial-jurisdictional scales*. From top to bottom, each level of governance covers small geographic spaces (domain). However, for different countries and provinces or cities, their approaches and their extent of involvement in port governance vary during different historical periods. When a state decentralizes its power and devolves tasks of port development and management to municipal government units, the city-level domain will claim more authority on port governance. At the same time, when national policies and port development worldwide become market-oriented and privatized, the number of stakeholders involved in port operations may increase. As a consequence, relationships between property rights will become more complicated. Similarly, when functional capabilities (logistical) of a port increase from handling single cargos to multiple functions, from merely focusing on importation to a combination of import, export, and re-export, and from simple highway cargo transportation to different modes of transportation, port administration will also become increasingly complicated.

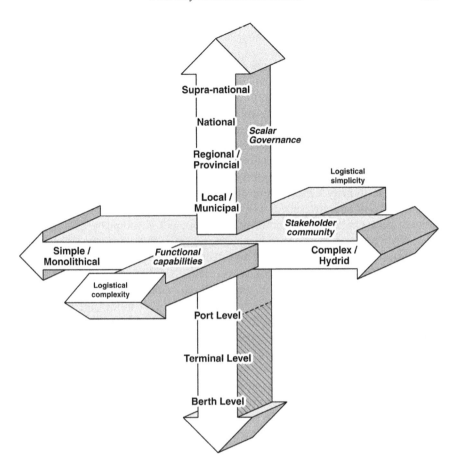

Figure 6.1   A Tri-Dimensional Model of Port Governance

Source: Wang, Olivier, and Ng, 2004

Diversified ways and degrees of governmental participation, complicated port property rights relationships, and varied port functional capabilities constitute the diversity of port governance. Therefore, while analyzing port-city relations in China at a specific period of time, we fully consider the characteristics of each axis and the dimension. These characteristics and the corresponding dimension include the following:

- Overall differences in port governance between China and other countries
- Different roles played by each domain for port governance, especially the role of the port authority

- Features and changes in the relationships between ports with regards property rights
- Differences in functional and business characteristics among a number of ports

From the perspective of port-city relationships, the above analytical approach of port governance is significant. First, given that each port differs in terms of property rights relationships and functional capabilities, even in the same country where each port city government is granted the same port administrative power, the effect of governance may still significantly vary. In other words, unless two ports share similar property rights relationships and functional capabilities, comparing them in terms of port governance will be difficult. Second, when the relationships between port administration authority and municipal government units change, especially when municipal government units prioritize their own port investments and constructions, the efficiency and the effect of port governance will be affected because port cities will regard port construction as a tool of inter-city competition and government departments will pay more attention to the development of business activities that are not directly related to the port in the port district for the purpose of improving economic efficiency in the entire jurisdictional area. Debrie Guovenal and Slack (2007) had analyzed the relationships between low-tier government units and ports in France and in Canada based on their respective devolution policies. Similarly, Shi (2003) and Mao (2005) emphasized on the devolution of port governance and that local government units at different levels will inevitably pay more attention to the development of ports than before. This devolution of port governance will be further discussed in the succeeding sections in the case studies of Chinese port cities.

6.2   Relationships Between Port Cities in Specific Areas

Neighboring port cities can belong to the same province, such as Dalian and Yingkou ports in the Liaoning province, or to different provinces, such as Rizhao port in Shang Dong and Lianyungang port in Jiangsu. From the perspective of a "province" as a specific area, the relationship between Dalian and Yingkou is overtly different from that between Rizhao and Lianyungang because the former two cities belong to the same province (Liaoning). Therefore, from the national perspective, a balanced development among provinces should be considered.

After 30 years of reform, of opening the country to international trade, and of developing the socialist market economy, China is regarded to have an executive-led economy in which all levels of the government demonstrate different approaches and abilities in restricting the development of a specific port city in a specific area and flexible methods that allow certain cities to circumvent conventional restrictions for development. Implementing restrictions and exhibiting flexibility are essential in port city governance because they are part of the spatial rescaling of port cities. In the next section, we will first elaborate on the present restrictions and capabilities in the development of port cities. A brief introduction to the theory of spatial rescaling is subsequently detailed. The succeeding chapter will demonstrate the possible effects of flexibility in specific case studies.

6.2.1   Major ways of restricting and balancing the development of different port cities

Favorable geographical positions and shorelines are prerequisites for the development of ports. The importance of ports and their functional capabilities have a positive correlation. A large-scale international port must have corresponding practices, security inspections, financial services, information services, land transportation services, and logistical services. In addition, an international airport is necessary to facilitate communication in its host city. Meanwhile, a small port intended for domestic trade requires only several highways for cargo transportation. Various ports not only differ based on their scale, but also on the examination and approval of high-level government departments. From the perspective of regional governance, any human factor that can affect the importance of a port in a region can be viewed as a restriction. For an executive-led economic system, government approval is the most effective method of restriction.

For an international trade port, the following items must be examined and approved (Table 6.2):

Table 6.2   Port-Related Classification of Approval Rights

Items for approval	Authority to approve			Remarks
	National	**Provincial**	**City-level**	
Scale and nature of ports	Ministry of Transportation or the State Council and related military authorities (see "Port Law" Article 9)			
Deep water berths	Ministry of Transportation			
Port land use	Ministry of Construction and Real Estate			
Bonded logistics zone	The State Council			
Export-processing zone	The State Council			

Table 6.2　　*Concluded*

Items for approval	Authority to approve			Remarks
	National	**Provincial**	**City-level**	
Port railway			Local government units above county level (see "Port Law" Article 21)	
Right of management of terminals			Port administrative departments and local departments of industry and commerce (see "Port Law" Article 22)	

Sources: "Port Law of the People's Republic of China" and "Examination and Approval Criteria and Procedures for the Establishment of the Export-processing Zone"

Some items for approval in Table 6.2 must be planned and programmed at the macro level to achieve an overall balanced development, such as the scale and the nature of the port, while others do not necessitate macroscopic planning (such as land use of the port).

From the perspective of the market, the cost of port construction and the economic efficiency of the port are not items for approval. However, the time required in requesting for approval and for the identification of the items to be approved will have a direct impact on the market position, potential for development, rivalry, and competitiveness of the port, unlike the case of many other countries. Apart from controlling port management rights, the government at all levels has other ways to facilitate or to prevent the development of ports toward market expectations.

Table 6.3　　**Chronology of Port Governance in China**

1949–1957	All major ports were managed by the Ministry of Transport
1958–1963	The management of all ports was decentralized to local government units
1964–1967	All ports were centralized to the central government for unified management, and the system of "port-shipping integration and regional governance" resumed
1968–1972	Port management was decentralized to local government units
1973–1983	All major ports were re-centralized to the central government for unified management

Table 6.3   *Concluded*

1984–2001	The state again carried out major reforms on the management system of major ports, except that Qinhuangdao port was still managed by the Ministry of Transport. The remaining 37 main ports along the coastal lines and along the Yangtze River, which were originally governed by the Ministry of Transport, were devolved to local government units. The central government started to implement the system of "central-local dual leadership with local-dominant management". In fiscal affairs, ports financed themselves and their revenues pay for their expenditures. Meanwhile, fixed-amount revenue, which will not change for several years, was turned over to the central government. Along the Yangtze River, ports and shipping were managed separately (port leaders and cadres were appointed by local government units, while port planning, construction, operation, and assets were managed by the central government).
2002–present	All ports, including the Qinhuangdao port and those under dual leadership, were decentralized to local government units. In principle, after decentralization, the ports were governed by the municipal government units of their host cities, and those that required governance from provincial government units were governed under the principle of "one port, one administration" with discretional management methods. After decentralization, port enterprises no longer exercised administrative functions but deepened their internal reforms and became legal entities accountable for their own losses and profits in accordance with the requirements in establishing modern enterprise systems. The current port plan and financial management system was reformed. The planned management by the central government was devolved to the local government units. For the financial system, the principle of "port self-financing and expenditures covered by revenues" changed to "separated revenues and expenditures". The approach of "port enterprises submitting a fixed amount of revenues" was abolished and business income taxes were levied according to regulations of the national tax revenue administration.

Port-related laws are another form of restriction. China has experienced several centralizations and decentralizations in terms of port governance since its establishment in 1949 (Table 6.3). The *Port Law of the People's Republic of China* or the "Port Law", which became effective on January 1, 2004, clarified uncertainties on legal and institutional affairs after the previously stated reforms, providing clear legal regulations for port governance. The following provisions are clear in this law:

1) Port administration responsibilities of the competent traffic authority of the State Council
Article 6 of the Port Law indicates that the competent traffic authority of the State Council is mandated to manage national port duties. Based on the current structure of the State Council, "the competent traffic authority" refers to the

Ministry of Transport that is in charge of national port duties and supervises the implementation of the Port Law throughout the country. According to this law and to other relevant legal mandate, as well as related regulations about the responsibilities, internal organization, and personnel arrangement of the Ministry of Transport, the responsibilities of the national port include (1) formulation of national port development strategies, guidelines, policies, and regulations, and supervision of their implementation according to the needs of the national economy and of social development, (2) planning the national port layout, supervising its implementation, and stating opinions about whether port layouts in provinces, autonomous regions, and in municipalities under the central authority are directly in line with the national port planning, (3) identification of ports in the country and overseeing their overall planning with government members of relevant provinces, autonomous regions, and municipalities directly under the central authority, as well as providing advice on regional major port planning, (4) establishment of standards for deep water coastlines of ports and examination and approval of the use of deep water coastlines for port infrastructure construction by the National Development and Reform Commission, (5) industrial administration on the port construction and identification of industrial technical standards and unified regulations on port construction and maintenance, (6) implementation of basic rules for port operation and safety, including regulations for handling dangerous cargos, (7) stipulation of conditions and approval of procedures to obtain qualifications for balancing businesses in ports, (8) management implementation by the collection of fees in ports, which are government-directed or government-fixed along with price administration departments under the State Council according to the 2011 Price Fixing Catalog of the State Development Planning Commission and of the State Council based on the Price Law, (9) national statistical work in accordance with the provisions of the Statistics Law and other government laws, (10) supervision and management of the collection and of the expenditure of port fees, (11) guidance for port administration duties of local port authorities (traffic authorities), (12) handling of administrative reconsiderations on dissatisfaction caused by port administrative actions of the Ministry of Transport or of provincial-level traffic authorities based on the Administrative Reconsideration Law, and (13) implementation of other port duties as stipulated by relevant laws and by the State Council.

2) Port administration responsibilities of local government units
The responsibilities of local government units regarding port administration are stipulated in Articles 6 (2) and (3) of the Port Law. "Local government units" refer to provincial, municipal, and county-level government units.

1. According to paragraph 2 of Article 6 of the Port Law, the management of ports by a local government in its administrative division will be determined according to the provisions of the port administrative system of the State Council. These provisions include those previously formulated by the

State Council (such as *Opinions on Deepening the Administration System Reform of Ports Directly under the Central Government or under Dual Leadership* by the General Office of the State Council with permission from the State Council) and new ones that are intended to further improve the port administration system.

2. Although *Opinions on Deepening the Administration System Reform of Ports Directly under the Central Government or under Dual Leadership* (referred as "Opinions") contain regulations on governing ports that are under the direct leadership of the central government or under dual leadership, the defined basic principles can also be applied to ports originally governed by local government units. "Opinions" stipulates two port administration system reform principles. The first principle is to separate enterprises from government management in port administration. Administrative functions shall be exercised by government departments according to law, while enterprises shall no longer be responsible for port management. The second principle is that ports shall be managed by corresponding municipal government units in their host cities. Ports that must be managed by provincial government units shall be led with the principle of "one port, one administration". Provincial government units will have the freedom to decide their management strategies. The number of ports that must be governed by provincial government units will be small, and the majority of the ports will be managed by local government units at both the city and the county levels. Questions whether or not some ports in the provincial administrative areas must be managed by provincial government units and which ports under the management of city-level government units must be governed by provincial or county-level government units will be answered by provincial government units according to the principles stipulated by the State Council.

3. Regardless of whether a port is governed by a city- or a county-level government or by a provincial government, the governing people shall designate one department to be responsible for port administration, as stated in paragraph 3 of this article. This department can comprise members of a traffic authority or a special department formed based on specific local conditions. According to Article 64 of the *Organic Law of the Local People's Congress and Local People's Governments of the People's Republic of China*, the formation of a new dedicated department shall be reported by the host government to a higher level of government for approval and to a standing committee of the People's Congress for record keeping.

4. Given that ports, in principle, should be governed by municipal government units of host cities, specific provisions about administrative responsibilities of provincial or autonomous region-level traffic authorities have not been mentioned because the management of ports are not under the direct leadership of provincial government units. The original draft of this law

has stipulated that traffic authorities of provincial government units or of government units of autonomous regions shall provide administrative guidance for port administration in their respective boundaries. Critics argue that the definition of "administrative guidance" in the original draft of this law is unclear, and traffic authorities, as competent individuals responsible for the transportation industry, should handle port governance, with the port as part of the transportation industry. Regarding functions of provincial and autonomous region-level government transport authorities, provincial government units or government units of autonomous regions should formulate specific policies based on local rules and regulations, on regulations of provincial government units or government units of autonomous regions, and on port administration principles formulated by the State Council.

Many problems appeared when the Port Law was first implemented. For example, the governing principle of "one city, one port" is controversial. According to the current administrative division in China, "one city, one port" refers to the idea that a prefecture-level city (or county-level city) establishes port authority to carry out administrative governance to the port industry in the entire city. Cross-regional operation of ports has been in conflict with the Port Law. For example, with the approval of the State Council and under the national strategy of building the "Shanghai International Shipping Center", Shanghai made a breakthrough by incorporating the Yangshan port area, which belongs to Zhoushan Islands in the Zhejiang province, into the governance boundary of the Shanghai port. Shanghai adopted the special port governance model "Shanghai + Yangshan", resulting in the emergence of the phenomenon "One city with two ports" or "one port shared by two cities". At present, the principle of "one city, one port" can no longer adapt to new port development trends or is at least insufficient for cross-regional development of ports. Thus, the current port administration system must be reformed to adapt to newly emerging cross-regional port development trend. This trend has reminded us of a special issue that port cities can engage in different flexible paths to circumvent previously mentioned restrictions, altering the development prospects of their ports.

6.2.2   Ways of circumvention

The following are common ways of circumvention to accelerate port development:

1. Merge ports to form a new port governance system
2. Merge several cities to form a new administrative unit
3. Form a cross-city port union through port enterprises (such as publicly listed port corporations)
4. Alter port functions and plan by introducing large port-related industries

These strategies usually render original laws and regulations inapplicable, and restrictions are circumvented. For example, when a port merges with another, the newly merged port is transformed into something that is not clearly defined in the laws and regulations, enabling local government units to benefit from the loopholes.

Government behaviors that adjust administrative boundaries and management systems all belong to governance activities of government units. Over the past 40 years, China has witnessed a number of these governance activities. To promote and to adapt to an export-oriented economy, China has made continuous breakthroughs in reforming and in developing its port and urban systems in a relatively short period of time. These reforms result in new development models for port cities. The common feature of these models is the recognition of port development as part of the development of cities or regions. In the next chapter, we will further interpret these models by applying them in specific case studies (Table 6.4).

Table 6.4   Various Methods of Circumvention in Port Governance

Ways of Circumvention	Cases	Details
(1)	Chongqing port	Chongqing, Wanzhou, and Fuling ports merged by a free overall transfer. Capital, creditors' rights, liabilities, and personnel were all transferred to the Chongqing Port Logistics Group. Wanzhou and Fuling ports became subsidiary companies of the Chongqing Port Logistics Group as a result of independent accounting.
(2)	Xiamen port	From January 1, 2006, three port areas originally governed by the Zhangzhou port in Xiamenwan and five existing port areas in the Xiamen port merged to form the new Xiamen Port. The original Xiamen and Zhangzhou port authorities rescinded, and Xiamen port authority was established to govern the port, shipping routes, and water transportation throughout Xiamen.
(2)	Meizhouwan port	In 2009, Xiuyu Port, originally governed by Putian in Meizhouwan, merged with Fujian and Xiaocuo ports that were governed by Quanzhou. Ports within the Meizhouwan area have similar names, institutions, planning methods, construction, governance, and services, subsequently establishing the Meizhouwan port authority.
(3)	Ports in Hebei	The Hebei port group was formed through the integration of port resources in the Qinhuangdao, Caofeidian, and in the Huanghua port areas based on the original Qinhuangdao Port Group. The port group is a provincial fully state-owned group setup based on the modern enterprise system, which has the functions of port construction, development, state-owned asset management, operations, investment, and financing.

Table 6.4   *Concluded*

Ways of Circumvention	Cases	Details
(3)	Qingdao port	In January of 2006, Qingdao and Weihai ports jointly invested 140 million RMB to build Container Terminal Co., Ltd., which marked the first step in the cross-regional development of Qingdao and in the cross-port integration in Shandong.
(3)	Rizhao port	In 2003, some of the enterprises in the Rizhao port authority and in the Lanshan port authority were rebuilt into the Rizhao Port (Group) Co., Ltd. On May 20, 2007, Qingdao and Rizhao ports invested jointly to build the Rizhao Riqing Terminal Co., Ltd.
(4)	Caofeidian port	The Shougang Group and Hebei Iron & Steel established a large steel production base in Caofeidian, while the Caofeidian port established a large terminal exclusively for ore.
(4)	Beibuwan port	Beibuwan focused on the construction of bases for petroleum, pulp and paper industry, energy, steel and aluminum processing, foodstuff processing, and marine industry. Beibuwan also plans to establish three professional transportation transit bases for bulk minerals, bulk grain, and containers to be concentrated in the Yuman and Qinzhouwan port areas.

Sources: Websites of the Fuling Port Association and of the Port and Shipping Authority in Fujian, www.xinhuanet.com, Weihai Port Group, China Network, Beibu Gulf ASEAN Economic Web

6.2.3   *Spatial rescaling*

Spatial rescaling or adjustment is a Neo-Marxist concept. Rescaling, as a spatial process, has been studied intensively by economic geographers for the a few decades since the concept was reemphasized as a fundamental concept by Harvey (1969) and has been referred to as a series of changes triggered by spatial rent-seeking activity of the capital, leading to the reallocation of investment in space at various geographical scales such as within a city, a region, or globally (Meentemeyer 1989). An example is the regional relocation of industries and of economic globalization. The relocation of capital results in corresponding scale adjustments of superstructures, such as political, institutional, and administrative organizations and governance. They are often represented by, but not limited to, relevant adjustments in terms of jurisdictional boundaries of spatial units to fulfill the requirements in producing a new space for the capital invested. As a result, new administrative authorities may be established at a new spatial scale or the scope of the original administrative management may be extended. Some methods of circumvention mentioned in the previous section can be regarded as spatial rescaling. However, spatial rescaling is not merely limited to the ports. Boned and development zones are two important tools used by local government units in many port cities in China during the rescaling process that affect port-city governance and efficiency as a whole.

As we have explained in Section 5 of Chapter 4, Tianjin is a good illustration of the interaction among bonded zone, development zone, and the port district, as well as the gradual transformation of the players in the international articulation space. This process has led to the creation of a new space in port cities in China. Section 6 of Chapter 4 emphasized on the administrative scope of Chinese port cities and regarded them as having the largest administrative scopes of port cities in the world (many of which were formed before the reform and the opening-up in 1978, whereas others were the result of adjustments of administrative boundaries, such as the incorporation of Panyu into Guangzhou), that enabled many ports to "leap forward".

Compared with other port cities, administrative and spatial arrangements of the previously mentioned zones in the formation of international articulation space in cities, such as Tianjin and Ningbo, were reasonable (see Figure 6.2). However, Zhuhai presented a relatively inappropriate administrative and spatial arrangement of the zones (see Figure 6.3). Two reasons cause such an awkward arrangement in Zhuhai: the Free Trade Zone was designed to be closely located to the land border to Macau rather than to a port in the city; and then when noting that this was a wrong decision made, the city government did not have the corresponding jurisdictional power to relocate it, or even setting up a new one – in China, such a special-tax-policy zone always requires the approval from the Commission of Development and Reforms under the State Council of Central Government.

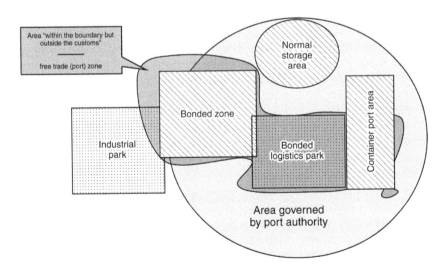

Figure 6.2   **Schematic View of Spatial and Administrative Relationships Between Terminals and Other Relevant Zones in Tianjin and Ningbo**

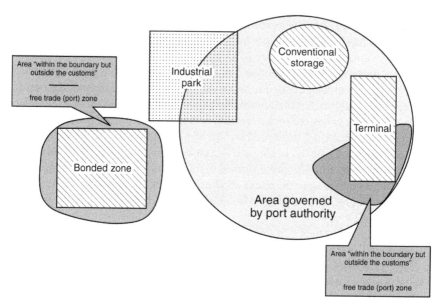

Figure 6.3   Schematic View of Spatial and Administrative Relationships Between Terminals and Other Relevant Zones in Zhuhai

6.3   Conclusion

The governance of Chinese ports and port cities has been experiencing rapid changes and adjustments. Compared with the adjustments in ports and their host cities in China over the past 30 years, countries in South America have only accomplished half of that of China. The foremost reform in port administration has been the decentralization of power from the national to the city government units. This decentralization provides a huge incentive to local government units by ensuring that ports, such as their instruments for inter-city competition, receive more investment from abroad and more export-processing industries will be established within their jurisdiction. Another noticeable change of ports is the frequent adjustment in the jurisdiction of port cities. To a large extent, adjustments of jurisdictional boundaries in port cities have created sufficient room for the expansion of ports and often caused shifts to shorelines with deeper drafts or at least increased the possibility of ports' expansion. As discussed in this chapter, the most interesting part of these reforms in port-city governance lies in the flexible and in the proactive means in which local (both city and provincial) government units have been using to achieve competitive power through territorial or jurisdictional rescaling and to get around the Port Law.

The means and the activities employed by local authorities demonstrate their real intentions behind port development. As we understand, the adjustment of city boundaries in China will increase, as well as the amount of resources that the cities will command. Therefore, as opportunities continue to be present, neighboring cities with ports will compete with other port cities rather than joining them. For example, the new Caofeidian port has new and unfamiliar districts in the city of Tangshan, creating a new port city rather than as a part of Tianjin port (as listed in Table 6.4). This new port-based district started in 2002 as a green-field project, strongly supported as a "dragon-head" (i.e., primary and leading) by the provincial government of Hebei. However, after an investment of more than 60 billion RMB, recent news[1] had reported that the district is in deep financial debt due to over-investment and to the redirection of development focus of the province to other cities, leaving Caofeidian port district as a ghost town. The man-made competition among port cities will continue, as well as the rescaling of port-city space, given the surprisingly easy change of city boundaries or the merging of two cities in this country. The next chapter will further discuss the competition and the cooperation among port cities from a regional perspective.

---

1   "Caifeidian, Tangshan, appears a financial risk" (唐山曹妃甸资金链临断裂危险), a new report by Liu Yuhai, 20130-52-5 accessed on June 5, 2013 at http://finance. qq.com/a/20130525/001268.htm.

Chapter 7

Relations Among Ports and Port Cities in Specific Regions

The previous chapter explained the influence of government behaviors on ports and port cities and the market response from the perspective of port governance. In this chapter, we will compare two cases from two different regions to examine varying development patterns as a result of the influence of government behaviors on China's regional port systems, on port-city relationship, and on other possible outcomes.

7.1 Competition, Division of Labor, and Cooperation Among Ports in the Greater Pearl River Delta (GPRD)

GPRD comprises the Pearl River Delta (PRD) in Guangdong Province and two special administrative regions, namely, Hong Kong and Macau. GPRD and the Yangtze River Delta (YRD), which are separated by Shanghai, are not only the two most economically developed regions, but are also areas that contain the most important ports and the highest volume of foreign and domestic trade in the country. However, GPRD and YRD differ in terms of economic and physical geographies and administrative systems.

1. Geographical scope. The area of YRD is approximately 50,000 square kilometers[1], while GPRD covers approximately 10,000 square kilometers[2].

---

1 According to "The Yangtze River Delta Regional Plan" approved by the State Council of People's Republic of China in May 2010, the YRD economic circle initially comprised 16 cities, namely, Shanghai, Hangzhou, Suzhou, Wuxi, Changzhou, Zhenjiang, Nanjing, Yangzhou, Taizhou (in Jiangsu Province), Nantong, Ningbo, Jiaxing, Huzhou, Shaoxing, Zhoushan, and Taizhou (in Zhejiang Province). The City of Taizhou (in Zhejiang Province) joined the delta as agreed in the fourth Meeting of the Forum of Economic Coordination for YRD Cities in August 2003. In the first edition of *Procedures to Become a Member of "The Forum of Economic Coordination for YRD Cities" (draft proposal)*, apart from the 16 existing member cities, Yancheng of Jiangsu Province, Wenzhou and Jinhua of Zhejiang Province, Ma'anshan of Anhui Province, Wuhu, Chuzhou, Xuancheng, Chaohu, and Tongling are also incorporated in the YRD city cluster. In 2004, Hefei and Ma'anshan of Anhui Province, Jinhua and Quzhou of Zhejiang Province, and Yancheng and Huan'an of Jiangsu applied for membership in the Forum. In 2008, as reported in *The State Council's Guiding Opinions on Further Promoting the Reform, Opening up, and Social-Economic Development of the Yangtze River Delta Region*, YRD included Shanghai, Jiangsu Province, and part of Zhejiang Province.

2 Located in Guangdong Province, PRD includes Huizhou of Guangdong Province (its urban area only, Huiyang, Huidong, and Boluo), Shenzhen, Dongguan, Guangzhou,

2. Administrative system. GPRD comprises two special administrative regions, namely, Hong Kong and Macau. Unlike YRD, port development in GPRD is neither equally under the governance of the Ministry of Transport nor is subject to the Port Law of the People's Republic of China (PRC) because of the "one country, two systems" principle that remains unchanged for 50 years until 2047. Meanwhile, Hong Kong is considered a free trading port outside PRC customs. Therefore, cargos shipped by coastal lines between Hong Kong and any port or city in Guangdong are categorized as foreign trade, which is completely different from the relation between any pair of cities in YRD and a port such as Ningbo or Yangshan port in Shanghai. The latter situation is similar to that between two PRD cities or ports or with any other mainland port, given that the shipments are regarded as domestic. Under the current regulatory environment in China, the cabotage policy prohibits foreign shipping lines from picking up containers from one mainland port and then delivering them to another mainland port. Consequently, the Hong Kong Port has an advantage in handling transshipments that a number of shipping lines prefer to conduct in the region, regardless of purpose such as interlining optimization.

3. Water system. The inland water system of the Pearl River is a dense and widespread network. The West River water system, which is part of the Pearl River, covers all cities in the western part of the Delta, while the middle and the eastern parts contain major ports. The relation between the Yangtze River and YRD is completely different. Many large inland ports of the Yangtze River, such as Wuhan, are not in YRD and are located much farther up and inland to the west. Despite the physical barriers and the need for improvement on the water draft of the middle and the upper streams of Yangtze, YRD ports, particularly in Shanghai, hold considerable potential. By contrast, given the relatively closer distance to various seaports in the region, cities in PRD have relied heavily on trucking, thus additional bridges crossing various branches of the Pearl River are constructed, thereby causing difficulty for inland water to flow from PRD into the river.

4. History of international trade gateway development. In GPRD, Guangzhou (Canton) and Hong Kong are two international trade ports that started at different points in history. In modern periods, Hong Kong replaced Guangzhou and became the largest maritime gateway in southern China. In YRD, although Ningbo and Shanghai once competed for superiority, Ningbo businessmen preferred to transact in Shanghai. Hence, Shanghai

---

Zhongshan, Zhuhai, Foshan, Jiangmen, Zhaoqing, and two special administrative regions, namely, Hong Kong and Macau. PRD is divided into GPRD and a smaller PRD. The smaller PRD does not include Hong Kong and Macau, unlike GPRD. GPRD is more familiar worldwide, whereas the smaller PRD is only known domestically. The PRD economic circle includes the nine local-level cities, while the GPRD city cluster includes Hong Kong and Macau.

is the number one port in eastern China during the entire twentieth century until presently. The dominant role of Shanghai in port businesses as the primary metropolis and trade hub in the region has never been challenged even when Nanjing became the capital city of PRC in the early 1910s.

With the special administrative system and the history of trade development, after China carried out its reform and its opening-up policy in 1978, GPRD witnessed unprecedented development in ports, manufacturing, and international trade. Many articles and discussions had been dedicated to the competition and cooperation among ports in Shenzhen, Hong Kong, and Guangzhou (Zheng Tianxiang 2005; Wang 1998; Song 2004; Wang and Olivier 2007). This chapter will not repeat the discussions in these articles but will rather focus on the answers to the following two questions:

1. What are the bases of competition among the three hub ports of Guangzhou, Hong Kong, and Shenzhen and how do they complement one another?
2. Geographically, what is the role of ports in a region and their corresponding transport systems in the formation and development of a cluster of mega cities?

The first question is on the market essence of the development of each port in the region, while the second is on the effect of various means of transportation on the growth of a specific city cluster in a region.

7.1.1   Competition and complementarity among Guangzhou, Hong Kong, and Shenzhen

From historical and geographical perspectives, for more than 100 years from the second half of the nineteenth century to the 1980s, the entire South China had only two real large-scale gateway ports for international trade, namely, Guangzhou and Hong Kong. The emergence of Shenzhen as a special economic zone (SEZ) and as a port city was the result of a specific historical event (China's reform and opening-up) and the establishment of a special zone (economic zone) at a special place (bordered by Hong Kong) during the past three decades. During these years, Shenzhen transformed from a small town with less than 100,000 people into a metropolis with a population of more than 10 million. This rapid development is better attributed to the policy of "one country, two systems" in Hong Kong and the label of SEZ instead of spontaneous urbanization. The different policies in Hong Kong, Shenzhen, and in other parts of the Mainland (especially other cities in PRD), the favorable infrastructures and institutional system of Hong Kong as an international free port, and the flexible policies and cheap land and labor costs have paved the way to the northward relocation and rapid expansion of Hong Kong's manufacturing industries. This development among industries also gave birth to the so-called "shop in the front and factory at the back" production model in which Shenzhen and Dongguan (a prefecture-level city north to Shenzhen) served as the factory, while Hong Kong was the shop responsible for marketing and other financial- and logistic-related services. This production model that emerged between 1980 and 1990 has not only brought simultaneous prosperity between Hong Kong and Shenzhen but also led to the rapid rise of the Shenzhen Port.

In the early 1990s, the Chinese central government encouraged Hong Kong container terminal developers to invest in the Mainland to increase the country's total container handling capacity. At that time, the policy that guaranteed the rapid development of foreign trade was very important and timely. Companies that captured the development trend of the market and invested promptly in the Shenzhen Port were publicly listed companies such as Hutchison Whampoa, Kowloon Wharf, China Merchants Group, and COSCO Pacific that are either private Hong Kong-based or state-owned firms. Thus, significant attention should be given to the following aspects:

- Investment destinations of publicly listed companies were Yantian and Shekou port districts that are located, respectively, to the east and to the west of Shenzhen. Both port districts are adjacent to Hong Kong and are located at the two flanks of Shenzhen's downtown areas. At that time, Shenzhen was similar to an industrial city with these two port districts as its new and direct gateways.

- Yantian and Shekou port districts have adopted the same advanced equipment and management techniques as Kwai Chung Port in Hong Kong and are managed and operated by the same parent terminal operators such as Hutchison Port Holdings and Modern Terminals Limited under Kowloon Wharf. The two port districts did not only considerably increase the overall container handling capacity in the entire region (including Hong Kong's Kwai Chung Container Port), but also significantly reduced the costs of high land transportation fees charged by Hong Kong drivers and the expensive cargo handling fees in Hong Kong ports.

- When people regard Hong Kong and Shenzhen as two separate cities, they always consider their fierce competition with each other. However, both the Shenzhen Port Authority and the Hong Kong Port Development Council (the current Port Development Committee) depend on a "hands-off" stance or laissez-faire policy, which is different from most port management models in mainland China (Wang, Olivier, and Ng, 2004). Thus, the construction of Shenzhen's container terminals is a trust-like spatial expansion of Kwai Chung Port in Hong Kong. The trust-like spatial expansion refers to the emergence of a regional oligopoly in which several companies from one city invest in building and in operating the same business in a nearby city. Although policies of the Chinese government stipulate that mainland state-owned or quasi-state-owned enterprises should own shares in Shenzhen's container terminals, their presence is basically insignificant. In the perspective of port businesses, the competition between ports in Hong Kong and in Shenzhen more involves terminal operators than two cities (see Figure 7.1).

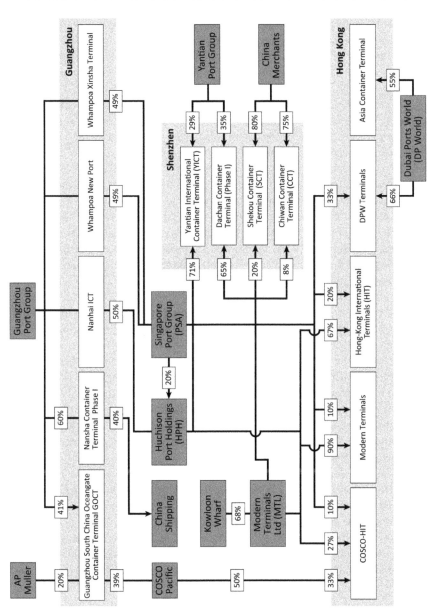

Legend: Dark grey box: the major shareholding companies, including port and shipping operators; Grey box: three major ports in GPRD, namely, Hong Kong, Shenzhen, and Guangzhou. White box and percentage: terminals with shareholding structure

Notes: In June 2005, Hutchison Whampoa sold 20% of the shares of HIT and 10% of the shares of COSCO-HIT to PSA for 925 million USD (approximately 7.49 billion RMB). In May 2006, Hutchison Whampoa sold 20% shares of HIT to Dubai Ports World. In 2007, PSA bought the stock ownership of the Nos. 3 and 8 terminals of Kwai Chung container terminals from NWS Holdings Limited (holding company of CSXWT). Since then, property ownership had become more complicated. The four largest port operators in the world (namely, HPH, PSA, Dubai Ports World, and China Merchants Group) and the Wharf Group in Hong Kong demonstrated an oligopoly in the region.

Source: Compiled by author, based on personal interviews and terminal website information

Figure 7.1   Shareholding Structure of Major Container Terminals in Greater Pearl River Delta (2012)

Thus, ports in Shenzhen are more reasonably considered as a spatial extension of Kwai Chung Port in Hong Kong rather than as two separate ports, which has already been explained with a model in Chapter 4. Hong Kong as a port has undergone two spatial expansions or relocations since its beginning. The first spatial shift of the Hong Kong Port was from the original Victoria Harbor to the Kwai Chung port district in 1972. In that shift, the substantial change was twofold, that is, an expansion in a new and deeper water location in a suburb and an establishment of a port district for containerized cargo. The second expansion occurred during the establishment of Yantian and Shekou port districts beyond the border of Hong Kong, to the east and to the west sides of Shenzhen, demonstrating a complete process of port regionalization (Notteboom 2007). The Dachanwan port district (another port district after Yantian and Shekou in Shenzhen) that started operation at the end of 2007 was merely a continuation of this regionalization process.

The case of Guangzhou is completely different. First, in terms of collaboration among terminal operators, Guangzhou Port has never formed any significant cooperation with ports in Hong Kong or in Shenzhen. Unlike Hong Kong and Shenzhen, Guangzhou is one of the eight "traditional Chinese ports" during and before the period of command and closed economy from 1949 to 1978. Therefore, Guangzhou has always been considered as one of the key ports under the Ministry of Transport in China. Moreover, reforms in the Guangzhou Port had always corresponded to the national reform and opening-up policies. In other words, since the founding of PRC in 1949, this capital city port had always been equally governed by governments and by enterprises under a planned economy until the mid-1990s when the national port system reformed administratively. Second, as one of the five most important hub ports (with Shanghai, Qingdao, Dalian, and Tianjin) in the country before 1978, Guangzhou primarily functions as the most important logistics center in South China, as well as the distribution center of material supplies from external sources to the Guangdong Province, especially PRD. At that time, Hong Kong was recognized as the center for foreign trade, while Guangzhou was a gateway for domestic trade. In the 1990s, Shenzhen emerged as a brand new port. Meanwhile, Guangzhou Port introduced PSA (Singapore Port Group) for the development of the container business. Later in 2000, the county-level city of Panyu was annexed into Guangzhou's administrative boundary. Following this geographical rescaling was the construction of the coastal deepwater container port district that was 70 km south to the original city of Guangzhou, replacing the expansion plan along the Pearl River (Zhujiang).

Initially, the Nansha port district aimed at domestic cargo only when its first phase commenced operation in 2004. This phase transformed the entire Guangzhou Port to a comprehensive and balanced hub port in the region for all cargo types. To illustrate the characteristics of the multi-function port, Table 7.1 lists the percentage of various types of cargo and that of foreign trade in Guangzhou Port in 2007. A considerable number of bulk cargos in the Guangzhou Port, such as coal, were not only supplied to Guangzhou, but to the entire Guangdong Province as well. Meanwhile, ports in Hong Kong and in Shenzhen provided regional and international container shipping services, and bulk cargos are basically inbound for local use.

Table 7.1   Composition of Cargos Handled in Guangzhou Port in 2007

Item	Guangzhou Port		
	Total	Import	Export
Total throughput:	34,325.08	21,297.06	13,028.02
Foreign trade	8,050.95	4,792.96	3,257.99
Domestic trade	26,274.13	16,504.1	9,770.03
Among which: coastal areas	26,274.13	16,504.1	9,770.03
Types of Cargo:			
Coal and related products	8,721.12	6,074.26	2,646.86
Oil, gas, and related products	3,206.7	1,692.26	1,514.44
Metallic minerals	456.79	421.64	35.15
Steel and iron	1,135.06	937.31	197.75
Mineral-building materials	4,587.82	3,245.4	1,342.42
Cement	12.17	5.42	6.75
Timber	218.05	144.12	73.93
Non-metal	207.19	164.57	42.62
Chemical fertilizers and pesticides	55.84	37.46	18.38
Salt	31.34	28.91	2.43
Grain	738.14	533.19	204.95
Machinery, equipment, and electric appliances	2,136.69	1,027.71	1,108.98
Chemical raw materials and related products	458.27	344.36	113.91
Non-ferrous metals	37.93	16.19	21.74
Light industry, medical products	566.32	244.04	322.28
Agriculture, forestry, animal husbandry, and fishery products	262.37	222.49	39.88
Other kinds of cargo	11,493.28	6,157.73	5,335.55
Total number of containers (10,000 containers)	**925.88**	**448.33**	**477.55**
Total weight (10,000 tons)	**12,329.95**	**6,492.1**	**5,837.85**
Domestic subtotal:			
1. Number of containers (10,000 containers)	593.41	282.93	310.48
2. Weight (10,000 tons)	8,939.1	4,796.13	4,142.97
International subtotal:			
1. Number of containers (10,000 containers)	332.47	165.4	167.07
Among which: via Hong Kong	190.17	85.3	104.87
2. Weight (10,000 tons)	3,390.85	1,695.97	1,694.88
Among which: via Hong Kong	2,159.98	992.45	1,167.53

Source: Guangzhou Port Authority

The development of container transport and of the degree of containerization is undoubtedly correlated to the industrial structure of a region. Export-processing and manufacturing industries that mainly focus on final consumer goods in PRD are the major client of container transport service. Ninety-four per cent of the cargos for import and for export in Hong Kong are shipped in containers. If a port mainly focuses on containers and not on raw materials and fuels (such as coal and grains) for its host region's consumption, the port will not consider converting its costly land resources and shorelines into a bulk cargo shipping center. Therefore, if the ports in Guangzhou develop larger container terminals and more ocean container shipping lines, they will be in fierce competition with ports in Hong Kong. However, Hong Kong (or Shenzhen) will not compete with Guangzhou in terms of shipment of bulk cargos and of basic raw materials for regional consumption.

From the above explanation about competition and complementarity among ports in Guangzhou, Hong Kong, and Shenzhen, the entry of the supply of regional container shipping services at the final stage of the S-curve can be understood, when Guangzhou and Shenzhen further developed Nansha and Dachanwan container port districts, respectively. This happened during the 2008 global financial crisis that led to large-scale business recession in container-reliant ports in Shenzhen and in Hong Kong, resulting in the oversupply of container shipping services in this region.

The ports in Shenzhen and in Hong Kong share the same port system and are managed by the same terminal developers that mainly supply container shipping services, whereas the Guangzhou Port is a relatively independent and multi-functional regional hub. Thus, how can the influences of these ports on the whole region or city cluster be evaluated and identified?

7.1.2   *Relationship between the entire port system and the development of the city cluster*

With regard the relationships between the entire port system and the development of the city cluster, three aspects should be analyzed.

The first aspect is about the expansion of the ports toward deeper waters and its role in the development of clusters of large ports and port cities. At the mouth of the Pearl River, the earth's rotation and the direction of water flow provide the river banks with a special characteristic. The shorelines facing east are shallow and silty, whereas the shorelines facing west are composed of rocks. Thus, shoreline conditions near the Qianhai new area in Shenzhen are better than those in Guangzhou, Zhongshan, and in Zhuhai. The Yantian Port in Shenzhen has an even better deep-water shoreline conditions because the port is free from silt at the mouth of the Pearl River. As previously mentioned in Chapter 4, the Guangzhou Port was relocated from the city center to Whampoa and finally to Nansha where port waters are increasingly deeper. Two features are significant in this process. First, Nansha, where sediments have settled for more than 100 years, is relatively new and is affected by increasing population and environmental concerns as the city center. Therefore, Nansha has immense potential to become a new industrial city. Second, except for ferry and passenger services, other port services will gradually decline in the original port district, transforming the focus from the landscape value to waterfront development (refer to theories in Chapter 1). However, given the very long shorelines of the Pearl River in the Guangzhou territory, the complicated port functions, and the important economic role of Guangzhou as a logistics center in the entire province, port authorities will not immediately abandon the Whampoa port district. The movement of businesses at Whampoa port district toward Nansha largely depends on the pace of the urban sprawl in Guangzhou and on the evolution and economic development of the Nansha district rather than on the port alone.

The situation in Shenzhen is different from that in Guangzhou. Ports in Shenzhen are an extension of the port in Hong Kong, thus Shenzhen has a dualistic urban space comprising the spatial extension of Hong Kong and the metropolitan areas of Shenzhen. Although the two city centers are apart from each other, both Hong Kong and Shenzhen have or tend to have their central business district (CBD) located somewhere in an area with the best waterfront. For Shenzhen, its existing CBD is in Futian district, which is newly built, but is not ideal. A recent development indicates the new plan of the Shenzhen government to build the second CBD in Qianhai, which has been approved by the Central Government as a "New District" with more favorable policy environment than SEZs in general. Qianhai is located in a reclaimed land area along the coastline and is between Dachan and Shekou port districts, thus causing a serious spatial conflict in terms of land use. The situation in Qianhai is unique compared with the usual practice worldwide, as illustrated by Bird's model in Chapter 4. For Shenzhen, a young port city with rapid urbanization and industrialization, a shift in development focus toward more high-level businesses, such as the financial market in Qianhai, or more growth in port throughput based on more manufacturing and export-oriented production remains unclear. Besides, the most advanced financial center in Asia is merely across the border in Hong Kong.

The second aspect is about how the entire PRD sustains its three ports, specifically the relation of the regional transport system and the ports to the cluster of cities. Both hub ports (including Hong Kong) in PRD and other major coastal ports in the Mainland have regarded highways as their primary channel in transporting the feedering and the collection of cargo. As railway transportation comprises only a very small proportion because of hinterland characteristics, cargos usually come from a region within 200 km from the three major ports. The percentage of river transport has also been small in comparison with that in major European port cities. Considering the infrastructure investment structure of Guangdong Province (Figure 7.2 and 7.3), this situation is generally consistent in recent years. The percentage of cargos transported by river shipping and by railways is minimal compared with that by highway transport. The percentage of cargos transported through highways will be larger if container shipping is

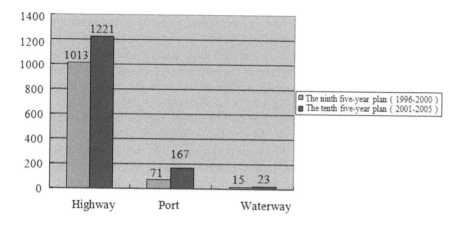

Figure 7.2   Traffic infrastructure investment in Guangdong Province: 1996 to 2005

Source: http://www.gdcd.gov.cn

considered. One of the reasons contributing to this increase is the growth of private export-processing industry clusters in the east part of PRD along with the development of deepwater container terminals in Hong Kong and in Shenzhen. Factories of these enterprises first moved from Hong Kong to Shenzhen SEZ in the early 1980s, and then further expanded to Dongguan and Huizhou. At the

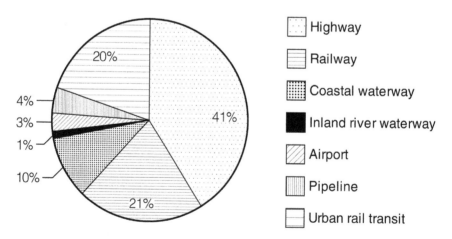

Figure 7.3   Percentage of investment in the overall transportation system of Guangdong Province during the eleventh five-year plan (%)

Source: http://www.gd.gov.cn

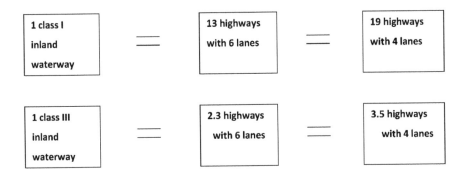

Figure 7.4   Capacity Comparison between inland waterways and highways

west bank of the Pearl River, branded enterprises, such as Midea (for electrical appliances), Galanz (for microwave ovens) in Shunde, and Gree (for air-conditioners) in Zhuhai, as well as famous domestic industrial clusters, such as the ceramics and lighting industry in Foshan and the furniture industry in Shunde and in Zhongshan, gradually entered the international market from their domestic foundations. These domestic market-oriented enterprises did not rely on ports in Shenzhen and in Hong Kong during the first one or two decades of their development.

Although transportation and openness are not the main causes for the varying market focus of industrial products at the east and the west banks of the Pearl River, the clustering of enterprises owned by Hong Kong and by Taiwanese businessmen at the east bank of the Pearl River is closely related to the accessibility of the highways that directly lead to international trading ports. Container trucks can make two round trips between factories at the east of Humen Bridge and Shenzhen or even at Kwai Chung Port in Hong Kong in a single day. The presence of highways has therefore significantly increased the efficiency of trucks and reduced road transport costs. For cities at the western part of PRD, trucks can make only one round trip to Shenzhen or to Hong Kong, consequently incurring higher transportation costs than those of cities at the eastern PRD. However, a cheaper means of transportation, via the river, is available in the western part of PRD.

In PRD, river transport has the most economic and environmental benefits, but is also the most easily overlooked. The water system of the Pearl River, especially its main stream, the Xijiang River, covers a large area. Cities, such as Foshan and Zhongshan, form a net-like structure along the Xijiang River, while other cities, namely, Shunde, Zhaoqing, Zhuhai, and Wuzhou in the Guangxi Province form a stripe-like structure.

According to the author's analysis, the transport capacity of a class I inland waterway is as large as that of 13 six-lane or 19 four-lane highways, while the transport capacity of a class III inland waterway can be as large as that of 2.3 six-lane or 3.5 four-lane highways (Figure 7.4). Table 7.2 exhibits the comparison of characteristics between river transport and other modes of transportation in GPRD.

Table 7.2   Characteristics of the Three Modes of Transportation in the Pearl River Delta

Characteristics	Inland waterway	Railway	Highway
Length	19,753 km	4,972 km	276,207 km
Cargos transported	Containerized Cargos Dry bulk cargos Liquid bulk cargos Dangerous cargos	Containerized cargos	All kinds of cargos
Transport scale	Large quantity of cargos, depending on the class of the waterway and the ship size	Depending on the length of the train	Depending on the maximum carrying capacity of the truck
Time reliability	Occasional jam	On schedule	Guaranteed delivery time
Reliability problems	Weather-related issues such as water level	Traffic congestion, waiting time	Congestion, accidents, weather (heavy rain, typhoons), crime
Safety	High	Middle	Below middle
Energy consumption/ pollution discharge	Lowest/lowest	Middle/pollution discharge decided by traction power of the train	High/high
Shipping cost	Lowest	Middle (with subsidy from the central government)	High

Source: Statistical Yearbook of Guangdong Province, 2008; Statistical Yearbook of Guangxi Zhuang Autonomous Region, 2008; PINE, 2004

The comparison between river and highway transport considers the following three aspects: time, cost, and convenience Table 7.3).

Table 7.3   **Distance and Cost of Highways and Waterways in the Pearl River Delta (2007)**

		Hong Kong		Nansha Port in Guangzhou		Shenzhen	
		Distance (km)	Cost (Yuan/ TEU)	Distance (km)	Cost (Yuan/ TEU)	Distance (km)	Cost (Yuan/ TEU)
Foshan	highway	200	3,400	65	455	170	1,190
	waterway	178	2,548	80	845	150	1,318
Zhaoqing	highway	263	1,675	183	1153	200	1,260
	waterway	228	1,864	148	631	200	1,003

Note: Waterway cost includes port loading and unloading costs as well as transport cost. Port loading and unloading costs are 1302 RMB/TEU in Hong Kong, 220.5 RMB/TEU in Nansha Port in Guangzhou, and 472.5 RMB/TEU in Shenzhen

In PRD, choosing river transport is more reasonable at the transport distance of more than 200 kilometers. With regard to time, the most important concern is not the speed of the river transport but the quantity of cargos in each shipment. Economic efficiency of river transport depends on the quantity size of cargos in each shipment. A ship with 3000 DWT can carry at least 150 forty-foot containers at one time. If the quantity of cargos is not large enough, some ships refuse to depart until they are sufficiently loaded. Thus, after all links such as production, logistics and multi-modal transport reach a certain scale, only then can river transport compete with highway transport. However, due to the low threshold of the river transport industry, related enterprises are generally small in scale and compete against each other despite being low-level businesses. In addition, as a result of limited understanding and the number of systemic problems in the past 30 years, inland waterways have not been sufficiently invested and managed and have been restricted by unreasonable technological regulations such as ship length and bridge clearance. Accordingly, no competitive inland waterways and ships or fleets have been developed. Consequently, only small ships (shorter than 50 meters without push boats), half-loaded ships, and shuttle-service ships that depart and return on the same day are used regularly to transport containers to Hong Kong and to Shenzhen.

This situation confines river transport of containers to places that are only several kilometers away from the river mouths. River transport networks, together with highway networks, constitute a port-centered containerized area with a radius of approximately 150 kilometers. The range of this area guarantees that vehicles or ships can make day trips and that next-day shipment schedules will not be held up. This is a common phenomenon in China's coastal areas, which is a reflection of the scope of China's foreign trade or as this paper labels as "the containerized

China". While trading for domestic consumption still uses traditional modes of transportation at a large extent (i.e., by train or by non-container trucks), distribution of products from the western part of PRD is still conducted through railway and highway networks.

The third aspect is about the influence of the gradual outward movement of ports in the logistics industry. This book has demonstrated the typical characteristics of the seaward movement of the Guangzhou Port from the downtown area in Chapter 4. After Guangzhou annexed Panyu in 2000, Guangzhou Port established container terminals instead of bulk cargo terminals in the Nansha port district. However, due to the large number of water-to-water transshipment activities with other small ports in PRD and the shortage of cargo sources or the limited consumer market, the international value-added logistics industry supported by container terminals did not emerge. Presently, in China, products are directly stored in containers and then exported to other countries. Relevant value-added logistics activities, such as distribution and post-packaging, are usually conducted at destination ports such as Rotterdam in Europe. The logistics industry in China still focuses on the comprehensive transportation and on the storage of break bulks, such as steel products, and bulk cargos, such as grain, using traditional modes of transportation. This kind of logistics services often cluster in bulk cargo terminals and in related railway stations, forming stripe-like areas along specific railways or highways, with lengths ranging from two kilometers to three kilometers. Examples of these areas are those around Whampoa port district in Guangzhou and the old Tanggu railway station in Tianjin. These logistics services are not environment-friendly and cause traffic congestion and chaos. Meanwhile, these areas are populated by a large number of informal settlers who are vulnerable to market changes and who move to other places once bulk cargo terminals or railway stations disappear due to economic contraction.

To achieve a value-added GDP growth, Nansha port district must proceed to at least one of the following directions: (1) shifting the focus of the port from domestic (container) trade to international trade, opening a competition with ports in Shenzhen and in Hong Kong, (2) relocating bulk cargo terminals from Whampoa by a special cargo railway line, and (3) establishing port-dependent industries such as shipbuilding and petrochemical plant. The third direction may cause high CO_2 emission in the region, which the Nansha port district has avoided for many years. Recent actions implemented by the Guangzhou government and by the Guangzhou Port have shown a development thrust along these three directions. The first direction is phase two of the Nansha port district with six berths built by the Guangzhou South China Oceangate Container Terminal (GOCT), which is a joint venture of Guangzhou Port Group, COSCO Pacific, and AP Müller (APM). The port started their operations since the end of September 2008, during the collapse of the world financial market. The second direction is a railway development plan that connects Nansha to the Guangzhou North railway cargo hub, which has received approval. Lastly, a huge shipbuilding yard that became operational in 2007.

Figure 7.5   Nansha Port District in Guangzhou

Among these developments, GOCT in which AP Müller is one of the stakeholders has brought a substantial change to the port's role. Maersk, a subsidiary of AP Müller, moved its South China hub from Hong Kong to Guangzhou and was followed by top-ranking shipping lines. As a result, 38 weekly worldwide service routes became operational by May 2013, and some 4.5 million TEU were handled in 2012.

The new role of the Nansha port district positively affected the formation of new port regionalization. Table 7.4 summarizes the features of this regional division of port functions in GPRD based on the author's 2013 surveys on the three ports. Unlike the situation in the Rotterdam-Antwerp ports in West Europe, the regionalization of port operation in GPRD went through a unique path of development in the past 40 years. The historical events are as follows:

- 1972. Kwai Chung container port in Hong Kong began to operate.
- 1980. Shenzhen Economic Zone was established after China announced its open-door policy and implemented economic reforms in 1978, resulting in the huge expansion and relocation of Hong Kong's manufacturing and export-processing industries to Shenzhen and to PRD in general.
- 1993. Yantian and Shekou port districts were established, transforming Shenzhen into a port city. The port was managed by Hong Kong terminal operating firms that include HPH and Modern Terminals Limited.
- 1997. Sovereignty over Hong Kong was returned to China, and the "one country, two systems" policy was implemented, maintaining Hong Kong's status as a free-market economy for 50 years.
- 2000. Panyu (including Nansha) was annexed by Guangzhou. The move of the Guangzhou Port to Nansha was planned.

- 2004. As a key step to the development of the Guangzhou Port from a river mouth port to a sea port, the Nansha port district was operationalized by the Guangzhou government, with a focus on domestic container shipments.
- 2008. GOCT began to operate, and Maersk decided to move its hub from Hong Kong to Nansha. The Dachan port district in Shenzhen also started operations.
- 2009. Shenzhen and Hong Kong experienced the first negative growth for the past 30 years, while Nansha maintained its positive growth due to its domestic focus.
- 2012. Nansha had more than half of its port throughput from international trade as a result of its unique market connection with the Middle East and Africa.
- 2013. Nansha township increased administratively as the Nansha New District. The district was awarded a favorable policy environment and higher level of autonomy by the Guangdong provincial government to encourage more foreign trade-based development.

At present, the division of port functions, with Hong Kong as the regional transshipment hub, Shenzhen as the main port in the east of PRD, and Guangzhou as the main port in the west of PRD, has been clearly established.

Table 7.4   Geographical Division of Functions Among Three Major Hub Ports in the Greater Pearl River Delta

Geographical Function	West PRD (origin or destination in west PRD including Guangzhou)	East PRD (origin or destination in East PRD)	Non-PRD (neither origin nor destination in PRD or Guangdong)
Container Cargo	Mainly by Guangzhou Port at Nansha	Mainly by Shenzhen and Hong Kong and secondarily by Guangzhou at Nansha	Dominated by Hong Kong
International	Shared by all ports except for Yantian in Shenzhen	Primarily by Shenzhen and secondarily by Hong Kong	Hong Kong
Domestic	Guangzhou, mainly at Nansha	Shenzhen and Guangzhou	None
Transshipment	Shared by Nansha, Shekou, and Hong Kong	Hong Kong and Shenzhen	International transshipments dominated by Hong Kong
Bulk Cargo	Dominated by Guangzhou, currently at Huangpu, but may move to Nansha	Dominated by Guangzhou	None

7.2   The Relation Among Beihai, Qinzhou, and Fangchenggang Ports in the Guangxi Province

Guangxi Zhuang Autonomous Region is a provincial-level region in southern China. In terms of overall economic strength or the throughput and total import and export trade volume of the ports, Guangxi emerged as the last or the second to the last among China's coastal provinces (see Table 7.4). Regarding scale and economic strength of individual ports or cities, Guangxi is far behind other coastal provinces (see Figure 5.3 for the international connectivity of Fangchenggang Port in Guangxi). By analyzing the relationships among the three ports in the coastal areas of Guangxi, we expect to present a relatively backward province in GPRD while demonstrating the development model of port cities in economically weak areas.

Among the three port cities in Guangxi, Fangchenggang Port is the first and the best-developed. This port expanded during the Sino-Vietnam War in the 1960s as a front support. The Fangchenggang Port has favorable harbor conditions and is equipped with a well-developed local transport system. In addition, like many strategically important ports such as La Havre in France, the port district of the Fangchenggang Port is large, while its host city is small. Therefore, the handling capacity of the Fangchenggang Port far exceeds local demand and the cargos handled by the port are not from or for the host city. Some leaders of the port are commuters who return to Nanning, the capital of Guangxi, during weekends.

Qinzhou, which is located between Nanning and Beibu Gulf (Gulf of Tonkin), remained as a small port and had not been included in any port planning until the early twenty-first century. However, Qinzhou Port unceasingly aimed to become a large port with a strong industrial base. Beihai is a coastal city with beautiful beaches and seascapes. After experiencing a serious real estate crisis in the late 1980s, Beihai realized the risk in the sole development of tourism. Hence, the city promoted instead local production and trade through ports and port-related industries. Qinzhou and subsequently, two other cities, formulated ambitious port development projects before 2006. Figure 7.6 and 7.7 exhibit the development projects of Qinzhou and of Fangchenggang respectively. The mid- and the long-term development projects of the three ports in terms of its planned total capacity exceeded the overall demand of the region. Meanwhile, each port overtly designed the waterways for their own interest rather than for the mutual interest of the three ports.

Table 7.5 Total Export Volume, Port Throughput, and GDP of Coastal Provinces Comprising Municipalities Directly Under the Central Government

	1987						1997						2007					
	Total export volume		Total port operations		GDP		Total export volume		Total port operations		GDP		Total export volume		Total port operations		GDP	
	1000 U.S. dollars	R	1000 tons	R	Billion RMB	R	1000 U.S. dollars	R	1000 tons	R	Billion RMB	R	1000 U.S. dollars	R	1000 tons	R	Billion RMB	R
Liaoning	3,788,000	3	47,260	4	147	4	8,897,000	7	88,690	5	349	6	35,325,000	7	414,920	5	1,102	7
Hebei	1,484,750	7	53,788	2	103	7	7,383,540	8	78,620	6	395	5	17,016,510	9	399,840	6	1,371	5
Tianjin	3,284,940	4	17,250	7	55	8	5,018,180	9	67,890	7	124	10	38,161,000	6	309,460	7	505	10
Shandong	2,975,820	5	45,282	5	175	2	13,085,430	4	113,770	4	665	3	75,243,740	5	575,470	2	2,597	2
Jiangsu	2,119,000	6	8,940	9	232	1	14,089,020	3	16,520	9	668	2	203,733,000	3	91,350	9	2,574	3
Shanghai	4,160,000	2	128,330	1	136	5	14,724,000	2	163,970	1	336	7	212,430,000	2	492,270	4	1,219	6
Zhejiang	1,234,060	8	39,953	6	130	6	11,112,810	6	121,870	3	464	4	76,803,530	4	574,390	3	1,878	4
Fujian	904,000	9	11,911	8	50	9	11,589,090	5	44,850	8	300	8	34,841,950	8	236,030	8	925	8
Guangdong	5,444,170	1	49,790	3	168	3	72,656,250	1	123,840	2	732	1	238,171,000	1	802,820	1	3,108	1
Guangxi	**543,310**	**10**	**1,590**	**10**	**44**	**10**	**2,382,660**	**10**	**11,780**	**10**	**202**	**9**	**2,877,410**	**10**	**71,920**	**10**	**596**	**9**

Note: R denotes ranking

Source: Statistical Yearbook of each province and municipality

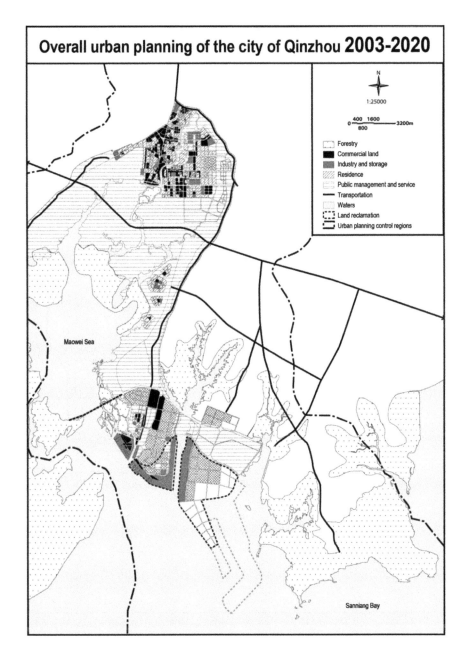

Figure 7.6   Qingzhou City Master Plan 2003 to 2020

Source: Author's research data

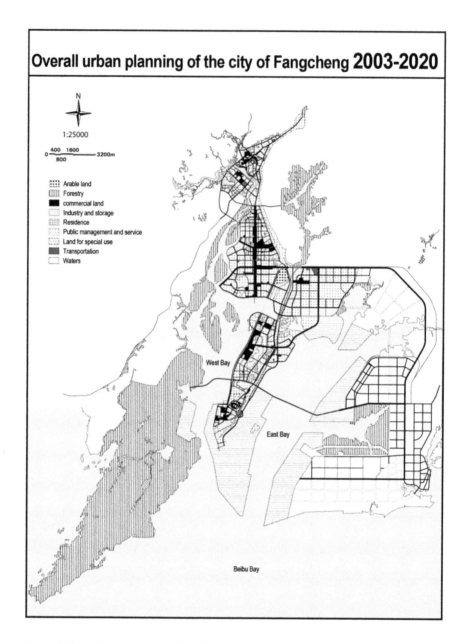

Figure 7.7   Fangchenggang City Master Plan 2003 to 2020

Source: Author's research data

The above port development projects and designs were formulated based on the consideration that the provincial government will select one of them to develop a signature port. However, things turned out otherwise. Fangchenggang exhibited good port conditions, while Qinzhou has a wide area of land and a large number of labor force. Besides, the aim of Beihai to become a well-developed port remained on a high level. If local demand sufficiently matches the capability of these ports, none of the projects will have adequate financial and human resources to independently develop a large port system. Therefore, the provincial government decided to establish the Beibu Gulf Coastal Economic Region (Figure 7.8), thereby preventing the competition among the three ports at a low level. By the end of December 2007, the National Development and Reform Commission reviewed and approved the *Development Plan of the Guangxi Beibu Gulf Coastal Economic Region (2006 to 2020)*. The plan was submitted to the State Council in early 2008 and was examined and approved in mid-January by Premier Wen Jiabao.[3] This economic zone covers an area of 42,500 square kilometers with a population of 12.55 million. The zone consists of the jurisdictions of "three ports and one city" (Beihai, Qinzhou, Fangchenggang, and Nanning).

According to the *Development Plan of the Guangxi Beibu Gulf Coastal Economic Region (2006 to 2020)*, the zone was "positioned" to become the western maritime corridor of China and as a key area in the conduct of cooperation with ASEAN countries. The so-called "positioning" of this economic zone in the development plan highlights the role of planners in a larger geographical scope. The approval from the State Council was a reflection of their recognition of this positioning. In the previous planned economy, all relevant ministries and government departments at different levels will have to comply strictly with this positioning. However, in today's open market economy, this positioning cannot guarantee anything except enabling the Guangxi government to allocate various resources to consolidate this positioning. Consolidation activities include the following:

- Policy. The government of Guangxi has submitted several authorization requests to the central government, including the establishment of a bonded port zone in Qinzhou, a comprehensive Sino-ASEAN bonded zone in the border city of Pingxiang, bonded logistics centers with export rebates in Nanning and in Fangchenggang, and the expansion of bonded logistics functions of the export-processing zone in Beihai.
- Administration in relation to the port hierarchical system. Apart from establishing the Beibu Gulf Coastal Economic Region, the original three port authorities were merged into the Beibu Gulf Port Authority that unified the situation at the regional level.

---

3   Refer to the news article in the website of Caijing (reporter Luo Wensheng): *"The State Council approved to set up Guangxi Beibu Gulf Economic Zone"*, January 1, 2008. http://www.caijing.com.cn/2008-01-22/100045979.html.

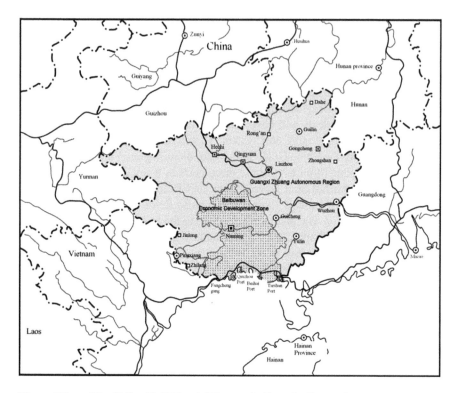

Figure 7.8   The Beibu Gulf Coastal Economic Zone in Guangxi

- Infrastructure. In addition to the construction of highways, a 50 km canal was to be built, which will directly connect the Nanning province to the sea.
- Logistics. The Guangxi government provided subsidies to shipping routes connecting Beibu Gulf Port to Hong Kong to offset the low appeal of a merged port (Beibu Gulf Port) at its initial stages of operation.

After the "economic zone" was established, favorable tax policies for various enterprises were enacted. These policies attracted enterprises such as CNPC, Wuhan Iron and Steel Corporation, Guangdong Nuclear Power Group, and Stora Enso, a Finnish paper manufacturer.

The Beibu Gulf Coastal Economic Region is an example of an entrepreneurial region in which the provincial government establishes and promotes a "place" that is as large as a group of four cities. This action may be interpreted as a strategic move of a weak province in a global and/or local competition with other actors who still function at the city level (e.g., Zhanjiang Port). In theory, this province is called an "entrepreneurial province" (Wang 2010). The similarity between an entrepreneurial province and an increasingly popular "entrepreneurial city" (Jessop and Sum 2000), locally and internationally, is the introduction of new types of "space" by the governments to facilitate participation in global competitions. The previously mentioned consolidation activities, as implemented by the government, are aimed at creating new types of "spaces" suitable for the clustering of certain industries and for new corridors for product flows. Thus, a new global logistics-based city-region is created. Guangxi is only one of the many cases of entrepreneurial regional development. Hebei is another example, in which without its capital focused on coastal resources, a new city-region of more than one million inhabitants built a new seaport on Caofeidian, an island 18 km away from the seashore.

Real economic benefits of positioning activities during the first few years are difficult to ascertain because the construction of infrastructure *per se* can generate GDP growth, a number of employment opportunities, and high throughput (e.g., the shipment of construction materials). Achieving real economic benefits depends on the region becoming a regional growth pole and a new gateway. Emulating this government-driven development model may be a challenge for countries with democratic political systems. Although the success of this government behavior is uncertain at present, provincial entrepreneurialism is found to be effective in avoiding overinvestment by municipal governments on ports in coastal cities of provinces.

7.3   Summary

The two port groups discussed in this chapter are extreme cases in China. PRD reflects a relatively more market-oriented system with special administrative circumstances, while the case of Beibu Gulf in Guangxi exemplifies a situation in which a proactive high-level government attempts to intercept the marketization of its ports. Other cases in China are in between these two extreme cases. An example is the development strategy of the west Taiwan Straits in Fujian Province. Although the provincial government intended to integrate Fuzhou and Xiamen Ports into one, both ports demonstrated obvious independence and a clear division of labor due to both ports having significant economic strength and administrative power. Similarly, Dalian and Yingkou Ports in Liaoning Province are both large-scale and multi-functional ports. They are also relatively independent and specialize in different cargos and hinterlands. Dalian caters more to the international market, while Yingkou focuses more on the domestic market. In this situation, the provincial government does not necessarily intervene in changing the entire port management system. Instead, the province can further encourage Dalian and Yingkou to take advantage of their merits in the business to attain exemplary performance.

The case of Guangxi shows the merging of three ports, while their respective host cities remained independent. This situation is not unique among ports in China's coastal areas. Other provinces, such as Zhejiang in which port grouping was implemented between Zhoushan and Ningbo Ports and Fujian (a port group consisting of Xiamen and Zhangzhou Ports), have witnessed the integration of ports. Although these port integrations are contributed by various reasons and are successful at various extents, a clear deviation from the traditional single management of the port either by the central government or by local governments is exhibited. Meanwhile, the abovementioned cases do not adhere to the rule that a port authority manages ports only in a single city, as imposed since 1994. Some cases have shown port companies making cross-city investments. Wuhan Port (an inland river port in the City of Wuhan) and other river ports along the Yangtze River, for example, were subjected to a joint operation by Shanghai International Port Corporation through cross-city port share-holding or share-controlling. Similar cases of port integration and joint operation indicate a new stage at which the market and the government simultaneously exert influences on port management. These new situations have different effects on ports and on their host cities, which is worthy for further exploration.

Another issue that deserves attention is the layout of large state-owned enterprises of the central government in port districts. These "central enterprises" or Yangqi include large machinery manufacturing enterprises of chemicals, oil, electric power, space industry, aviation, and railways. These enterprises are unlike many export-processing private enterprises, joint ventures, or foreign-funded enterprises in the following aspects:

- Their products are generally intended for the local market.
- The enterprises are generally excluded from tax rebates.
- The enterprises do not rely on container terminals, but develop and use their own user terminals. Therefore, they have special requirements for coastline conditions.
- The enterprises have more bargaining power in site selection negotiations with local governments because they have a higher administrative level.
- The enterprises are generally upstream or midstream enterprises in supply chains. Their ability to generate relevant downstream or derived enterprises is also one of their bargaining advantages.

The biggest risk that local governments may take in attracting central enterprises to their port cities is the failure of these enterprises to show up despite their expressed intentions to invest in the port city. This failure leads to the misuse of land areas originally allocated for them. This risk is not from the market but from the hidden and unpredictable process of "balancing" and bargaining both at the national and at the provincial levels. The risk may be reduced when provincial governments help local governments in capturing these enterprises. After obtaining these enterprises, their scale size at the point of investment remains unpredictable. On one hand, port cities frequently waste valuable lands in attracting these enterprises. On the other hand, a large number of central enterprises initiatively explore the area that the central government wants to develop (e.g., the Binhai New Area in the City of Tianjin). Therefore, the development of port cities in China is both market-oriented and command-based and planned by the government. Ironically, we observe the possibility of a relatively rational division of labor in port regionalization, as illustrated by the GPRD case in this chapter, though the actual paid cost was unknown. In Nansha of Guangzhou, for example, an 18 km deep-water channel must be maintained annually. The huge hidden cost is covered by the city government rather than by the port or the port users. Specific to this point, the Chinese government does not conduct cost-benefit analysis based purely on the project (port) operation, such as that by Western city governments or port authorities. This situation reflects a much closer relationship between the port and its host city in this country. From the regional perspective, the case of Guangxi demonstrated the same logic.

This case also demonstrates that in China where the political institutions remains 'exclusive' Acemoglu and Robinson (2012), the process of port regionalization may vary significantly from the conceptualized model described in Notteboom and Rodrigue (2005) where 1) the port development scale that matches the demand due to the economic globalization; and 2) it emphasizes the power of market forces represented by the major players in the global supply chains, largely, the shipping lines and global container terminal operators, as they have been penetrating regionally to the port ranges serving the major hinterlands and forelands. In the case of China, as argued here and other chapters in this book, it is the states, particularly the local governments concerned that play a critical role in altering the choice of these firms (market players) and eventually reshaping the places and spaces and hence the port-city relations.

Chapter 8
Multi-Layered Port-City Dynamics in China's Coastal Areas

8.1   Rapid Development of China's Coastal Port Cities Due to Globalization

One of the most important achievements of China's reform and opening up over the past 30 years is that China has become part of the global market economy, causing fundamental changes in the global economy as China's reform is a crucial link in the process of globalization. China's participation in the world system does not only bring material exchanges, political conflicts, and cultural clashes and penetrations, but also a brand new look to regions and cities in China involved in the process of globalization. These regions and cities are mainly concentrated in the coastal areas, which have been the actual birthplace of foreign trade and foreign exchange channels over the past two decades. This concentration has shown no decrease but rather an increase in recent years. Figure 8.1 shows that coastal provinces and municipalities whose trade volume accounted for more than 1 percent of the total national trade volume are all in the coastal areas. The total trade volume of the 10 coastal provinces and municipalities in Figure 8.1 accounted for more than 91 percent of the total foreign trade volume in China. The total trade volume of two provinces and one municipality in the Yangtze River Delta accounted for approximately 40 percent of the total trade volume while Guangdong accounted for 30 percent. Whether this rapid development in the coastal areas has been a success or has widened the development gap between different regions in China, summarizing and discussing the great changes in the coastal cities in the past two decades are worthwhile. This chapter focuses on port cities that have played a pivotal role in China's coastal development. This chapter discusses the center of evolution and the influences of multi-layered port-city relationships.

The first phenomenon that should be discussed is the increasing concentration of foreign trade in coastal areas. Ports' throughput in China has not only increased rapidly but has also shown clear dispersion to even more ports. As shown in Figure 8.2, although the total throughput of Shanghai Port increased from nearly 140 million tons in 1990 to 490 million tons in 2007, its proportion in the total national port throughput decreased from 29 percent to 12 percent. All ports listed in Figure 8.2 followed the same trend, except those that belong to the category of "other ports" whose total throughput increased from 8 percent to 16 percent. Figure 8.1 shows that this trend is not consistent with the trend of import and export distribution in China's coastal provinces and municipalities. Figure 8.1 also shows that since 1997, the proportion of the foreign trade volume of Guangdong Province compared

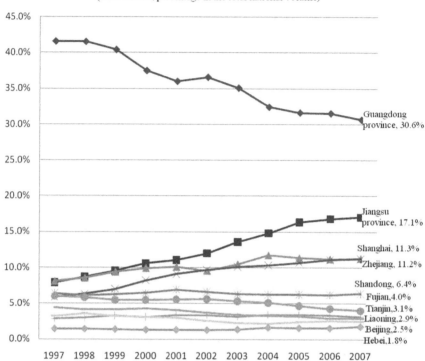

Top ten domestic destinations/origins of goods of Chinese import and export goods
(1997 - 2007, percentage in the total national volume)

Figure 8.1   Distribution of the Total Value of Goods Imported and Exported in China

Source: Related annual *Foreign Trade Statistics Yearbook*, China Customs

with that of the whole country has greatly declined, whereas that of the provinces and municipalities in the Yangtze River Delta has witnessed a marked increase.

This inconsistency can be explained by two reasons. First, the trade volume for domestic consumption was included in the ports' throughput. Second, in provinces with a large amount of foreign trade, such as Guangdong, Zhejiang, Fujian, Jiangsu, and Shandong, many small- and medium-sized port cities developed very quickly, which resulted in a large amount of repeated calculations of transshipped cargo.

8.2   China's Multi-Layered Coast-Centric Economic Development: Process, Characteristics, and Causes

China has undergone many steps of the coastal-centric process since the reform and opening-up of China in 1978. Coastal-centric means that more focus is given

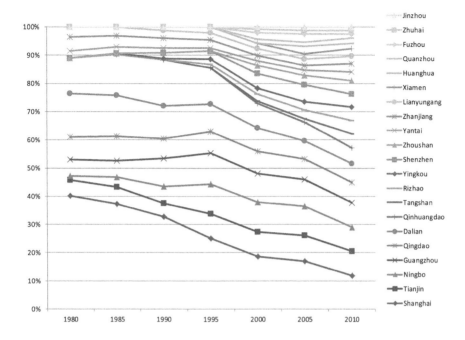

Figure 8.2   Major Chinese Coastal Ports: The Evolution of Port Growth and Throughput Distribution (1980–2010)

to coastal areas in the economic development of China. With the great vision from Deng Xiaoping's "reform" and "opening-up", the coastal-centric process in the past 25 years can be summed up in two ideas. "Reform" indicates power localization and economic marketization; "opening-up" indicates China's becoming part of the globalized economy and trade circle.

Specifically, the first or the earliest step in the coastal-centric process was to set up special economic zones and open 14 coastal cities in 1980. This step was the key in the country's move to develop the coastal areas. The establishment of many "sub-provincial cities with independent economic accounting" was a national policy to stimulate the development of coastal economies. At that time, Lu Dadao proposed the "inverted T-shaped" national development strategy with the coastal areas and Yangtze River as the axis (see p. 190 of *Chinese National Geography,* October, 2009 issue for reference). Did each coastal province apply the strategy of "coastal-centric development" consciously or unconsciously (market effect) afterward? We intended to analyze the question using urban GDP, city population, and built-up city area, but unfortunately we were able to collect relatively complete data on the GDP only. We compared five provinces, namely, Shandong, Zhejiang, Fujian, Guangdong, and Jiangsu (Table 8.1). Table 8.1 shows how the percentage of coastal cities' GDP in the provincial GDP changed from 1990 to 2004. Of the five provinces, Fujian and Guangdong underwent the greatest changes. Jiangsu is unique; a special column "Jiangsu 2" is dedicated to Jiangsu Province to show the GDP percentage of cities along the Yangtze River. According to Table 8.1, Jiangsu Province is river-centric rather than sea-centric, indicating that river ports have a better condition than coastal ports in the province. The coastal-centric process of Shandong Province stabilized in 1995 and that of Zhejiang stabilized in 1990.

Table 8.1   One of the Characteristics of the Coast-Centric Economic Development in Five Provinces Since 1990: Percentage of GDP

	1990 or 1991	1995	2000 or 2001	2004
Shandong	46.1%	50.5%	49.5%	49.4%
Zhejiang	81.0%	-	81.3%	82.3%
Fujian	70.0%	79.3%	83.0%	83.6%
Guangdong	72.3%	-	78.0%	79.4%
Jiangsu	21.4%	18.8%	18.8%	16.2%
Jiangsu 2	72.9%	77.1%	76.5%	79.3%

Source: Calculated based on the National Economy Statistics Yearbook of each province

Based on the types of industry and modes of transportation, national- and provincial-level coast-centric economic development for the past 25 years can be divided into two specific stages:

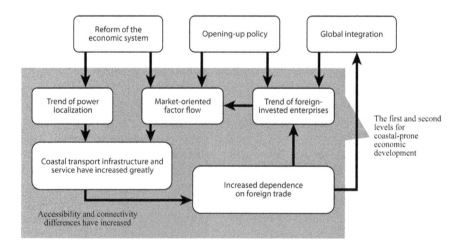

Figure 8.3   Reasons for the Multi-Layered Coast-Centric Economic Development in China

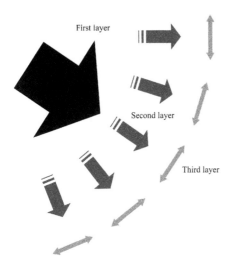

First layer

Second layer

Third layer

Figure 8.4　Three Layers of the Coast-Centric Economic Development in China

Note: (1) National Coast-centric Economic Development; (2) Provincial Coast-centric Economic Development; (3) City-level Coast-centric Economic Development (discussion on Section 8.6)

1. **Stage dominated by the export processing industry (1980–2000).** This stage gave birth to cities and towns that were considered "China's factories". It was an exogenous economic system aimed at the overseas market based on the many export processing enterprises that collaborated with hub and feeder ports (i.e., Hub and Spoke System).

2. **Stage in which the import of raw materials and fuels for industrial utilization increased significantly because of the increasing demand from industries for both export and domestic consumption (2001– present).** In this stage, many large coastal industrial zones dominated by heavy chemical enterprises emerged. It was an endogenous economic system aimed at domestic consumption. Large industries built dedicated terminals or port areas (Chapter 4).

These two stages, especially the recent 10 years (i.e., the second stage up to now), have witnessed two events. First, the import and export activities of goods in China have increasingly concentrated in the coastal areas (Table 8.2). Second, the growth rate of the total port throughput has further exceeded that of the national GDP (Table 8.3). These two events reflect the growth of an export-oriented and exogenous economy in China, that is, the essential reliance of Chinese economy on foreign markets, particularly of coastal China.

Table 8.2   Statistics on the Total Value of Chinese Import and Export Goods in Domestic Destinations/Origins of Goods (1998–2012)

Unit: billion US dollar

	1998	1999	2000	2001	2002	2003	2004	2005
National Total	324	361	474	510	621	851	1,155	1,422
Coastal provinces and cities	287	322	430	463	568	778	1,070	1,073
Inland	37	39	44	47	53	73	85	349
Percentage:								
Coastal provinces and cities	88.70%	89.20%	90.70%	90.90%	91.50%	91.40%	92.60%	75.5%
Inland	11.30%	10.80%	9.30%	9.10%	8.50%	8.60%	7.40%	24.5%

	2006	2007	2008	2009	2010	2011	2012
National Total	1,761	2,174	2,562	2,207	2,973	3,642	3,867
Coastal provinces and cities	1,623	1,989	2,318	2,018	2,710	3,276	3,436
Inland	138	185	243	189	263	367	430
Percentage:							
Coastal provinces and cities	92.2%	91.5%	90.5%	91.4%	91.2%	89.9%	88.9%
Inland	7.8%	8.5%	9.5%	8.6%	8.8%	10.1%	11.1%

Source: China Infobank
Note: The coastal provinces and cities include all coastal provinces and Beijing, Tianjin, and Shanghai.

Table 8.3   Comparison between the Increase in GDP and the Total Port Throughput

	1979–2004	1990–2004	1998–2004
Average annual growth rate of the national GDP	9.4%	9.3%	8.2%
Average annual growth rate of the total national ports	10.2%	11.4%	15.3%

Source: China Infobank

Some cities experienced these two stages, whereas others experienced only one. For example, Shanghai and Qingdao are cities where these two stages occurred simultaneously. Both cities not only built hub ports to serve as "China's factories" established in the first stage but also expanded and developed heavy industries and their corresponding dedicated terminals (e.g., Shanghai's Baoshan Iron and Steel Group built its dedicated ore terminal in the island of Majishan in the city of Zhoushan, and Qingdao built a dedicated oil terminal in its Huangdao port district). Not many cities experienced both stages at the same time. Generally, enterprises related to the import of raw materials, such as those in the oil and chemical industry and in the iron and steel industry, take a foothold in relatively marginal coastal cities such as Zhoushan, Zhuhai, Tangshan, and Zhanjiang. Conversely, export processing enterprises concentrate in cities with abundant labor, such as Shenzhen, Suzhou, and Tianjin, or in city clusters with feeder ports, such as Zhongshan, Quanzhou, and Wenzhou.

Different types of port cities have emerged because of various reasons. From the port's point of view, these reasons are as follows:

1. **Market force.** For example, the relocation of Capital Iron and Steel Group to Caofeidian, an island 18 km away from the seashore of Tangshan, was partly caused by the iron ores on which the company depended shifting from domestic to import sources. The shoe manufacturing industry in the city of Jinjiang in Fujian Province and the Taiwan-funded electronics industry in the Yangtze River Delta are the result of the capital's search for optimal manufacturing sites. These enterprises establish in coastal cities because they are world market businesses. As long as their markets do not change, these enterprises will not relocate themselves to inland areas even though labor cost in the coastal areas will increase.

2. **Natural endowment.** Water depth is the key. Ships utilized for international shipping today, especially trans-oceanic shipping, are all super ships (mainly oil tankers or ore carriers weighing 150,000 tons to 200,000 tons and 6,000 to 15,000 TEU container ships), regardless of the types of cargo (i.e., bulk cargos, oil, and containers). The usage of these ships resulted in the outdatedness of the old port areas in many major port cities, such as Guangzhou (including Guangzhou New Port and the Whampoa port area) and Shanghai (Port area on Jun-Gong Road), and old port districts in Tianjin and Qingdao. Newly built deepwater port areas leaped forward and away from the old port areas or the old urban centers. As deepwater shorelines are commonly found in different places (excluding Ningbo Port where deepwater shorelines are concentrated in one place), different port areas in a city can be 50 km to 150 km apart (e.g., in the city of Fuzhou, the Luoyuanwan port area is 160 km away from the Jiangyin port area). The scattering of port areas resulted in a port-centered development pattern. If the port's surrounding land areas are suitable for the development of heavy industry, establishing an industrial zone will be a good idea.

The surroundings areas will benefit not only from cheap land and rent but also from the reduction of the growing pressure from the environmental concern of petro-chemical or other heavy industries being adjacent to residential locations in urban and suburban districts.

3. **Role of the local government.** Over the past 30 years, Chinese city and municipal governments have regarded ports as a means of attracting foreign investment and a tool for competing. As discussed in the previous chapters, Chinese city and municipal governments have directly or indirectly subsidized their port development through various means. This subsidization is one of the reasons why ports in China have developed so rapidly. However, one must keep in mind that the ultimate goal of these governments is not port development itself. Understanding the incentives of the local government is the key to understanding their support. Two observable links may associate port performance with the goals of local government leaders. The first one is that the port may directly contribute to the GDP of the local economy, as the promotion and performance of the local government has been tightly linked to the GDP growth rate. The construction of a port itself can be counted into GDP! However, Chapter 4 proves that the association of port throughput with GDP growth has a log-linear relationship at the initial and take-off stages of port development. At the latter stage of port development, its contribution to the GDP may become smaller. The second link is that port growth to city leadership performance can be done by port ranking. I was personally told by the director of one port administrative bureau that only when the port becomes the world's number one would the bureau more seriously consider other objectives, such cost and the environment. It is odd but unfortunately true that port ranking among Chinese port has also been considered as sort of indicator of city leadership performance.

4. **Central planning.** The central government (i.e., the Ministry of Transportation (MOT) and the National Development and Reform Commission (NDRC)) has principles and preferences in port development in the country. Port areas such as Yantian in Shenzhen, Dayaowan in Dalian, and Beilun in Ningbo are all deepwater hub ports planned by the central government. However, local port constructions commonly circumvent these rules and principles. Most, if not all, port areas, such as the Yangshan port area built in Zhoushan of Zhejing Province by Shanghai, the Caofeidian port area under construction in Tangshan, and the oil terminal on the island of Daxie in Ningbo, among others, were planned and initiated by local authorities rather than by the central government or MOT. Today, an increasing number of these locally planned ports are being constructed. This phenomenon is caused by the fact that the economy develops faster than common planning. However, the major reason lies in the rapid marketization of ports as important nodes in the global supply chain. Recognizing the role of the market and the demand for global and

regional shipping, MOT and NDRC are still the central authority in control of key resources, such as deep-water seashore lines, through the Port Law and other regulations. In practice, MOT has to allocate room for the locals to seek development potential, whether real or not. Therefore, each province along the entire coast must be granted at least one regional "hub". For major economic regions such as the Yangtze River Delta and the Pearl River Delta, titles such as international shipping center are designated by the MOT or NDRC to the key ports, such as Shanghai, Qindao, Tianjin, and Dalian, so that these ports and associated host cities have room to grow and compete with each other and with ports in nearby Asia, such as Busan and Kaoshiong. For smaller port cities that will never be recognized as regional or major hub ports, such as Zhuhai in Guangdong, Qingzhou in Guangxi, and Lianyungang in Hebei, among others, they may attempt to lure individual firms in heavy industries, such as Petro-China, Sino Petro Chemical, and Baoshan Iron and Steel Group, to locate a new branch or production base to their port areas.

8.3 Competition and Division of Labor Between Chinese Port Cities

The competition between cities is essentially their contention for non-local capital. During the planned economic period, this kind of competition mainly took place in the annual meetings of the Planning Commissions at different levels. After the reforms and opening-up policy were implemented in 1978, market power became influential. Government interventions played an important role in this process, which is clearly reflected in the competition between port cities. In 1994, the State Council decentralized the power of port governance and operations to local governments and announced that the local governments should assume full responsibility of their own profits and losses. For the first time, port cities became the master of their ports. At the same time, ports began to be open to foreign investors. This event can be considered the starting point where local governments started to view ports as a means of competition. Foreign and other external investments (including capital from Hong Kong) took a few years to finally enter all of the major ports in China. Container terminals were built at an unprecedented pace, and the total throughput of many large ports (e.g., Ningbo, Shanghai, Shenzhen, Qingdao, and Xiamen) increased at an annual rate of 20 percent to 50 percent from 1995 to 2004. The ratio of the growth rate of total container throughput to the growth rate of local foreign trade volume and the growth rate of the GDP was 3: 2: 1, indicating the great contribution of ports to the local economy and foreign trade.

However, during this period of port expansion, debates on who should be the international shipping did not cease. The most typical case was the debate on how ports in the Yangtze River Delta could form a port system or a regionally "integrated port group" or "Zuhe Gang (组合港)" overseen by Shanghai. Despite the fierce debates, two facts were evident: (1) Neither the concept of the international shipping hub nor the concept of Zuhe Gang was able to stop or slow down the rapid expansion of ports. (2) Port cities did not specialize in certain areas according to their status or position planned by the State Council at that time. The construction and expansion of the Yangshan port area in Zhejiang Province for Shanghai, the completion of Hangzhou Bay Bridge, and the further expansion of major ports in Jiangsu and Zhejiang Provinces rendered the debate insignificant. The competition between port cities resulted in ambitious investments in the infrastructure and in the local transport networks in almost all ports, which brought huge economic benefits and paved the way for the relocation of the world factory to the Yangtze River Delta. Arguing who was the "hub" or the "center" was no longer important, as each port tried to expand further (Shanghai had the highest container throughput in the world, and Ningbo is among the world's top 10 ports in terms of container throughput and the third largest port in total throughput). Similarly, Shenzhen and Hong Kong in the Pearl River Delta became two of the top container ports in the world. Fortunately, no one could stop the emergence of several regional hub ports at that time. Otherwise, the regional environment and the economy would have greatly suffered.

From the national point of view, the competition between hub container ports in several major regions dominated by export processing industries (e.g., the Yangtze River Delta and the Pearl River Delta) can be considered the first stage of China's economic globalization. Two phenomena are notable in other coastal areas that are not so competitive in terms of their conditions for foreign trade, inland transport networks, urban infrastructure, financial services, and the Customs services, such as Guangxi Zhuang Autonomous Region, the west part of Guangdong Province, and Hebei Province. First, the provincial government directly participated in the development of specific ports or port cities, such as the Caofeidian port area in the city of Tangshan in Hebei Province. Second, small- and medium-sized ports tried to lure heavy chemical and industrial bases to their shores.

If this inter-port competition for the status of hub container port is viewed as the first level of competition between port cities in China, the second phenomenon reminds us of the second level of competition. At this level, deepwater coastlines and a broad range of land areas are the means of competition, and large-scale petroleum and chemical enterprises, power plants, shipbuilding factories, iron and steel manufacturers, and papermaking enterprises are what the port cities compete for. Local governments are the major competitors. For example, in Fujian Province, Quanzhou and Putian are cities located on two sides of the Meizhou Bay. They have always competed for the resources there and utilized the same resources, such as the natural deepwater bay, although this region is generally physically blocked by mountains to all inland provinces in Southeast China. Another example is Guangxi, where the three port cities of Fangchenggang, Qinzhou, and Beihai share the coastline of the Beibu Gulf. Each port city works for its own interests and has its own ambitious plan for the development of the port and port-surrounding industries (see the discussion in the previous chapter). For port cities, this level of competition and influence is different from the first level.

First, the requirements for establishing a port for heavy chemical and industrial bases are lower than the requirements for building a hub container port. In other words, as long as a city has a certain length of deepwater coastline, it will be able to lure heavy chemical and industrial enterprises to its shores, provided that the firms consider that region as their potential market and the local government can prove that such development will not cause any significant environmental consequences (although it may not always be the case). Therefore, more coastal cities will participate in the competition. Second, what port cities compete for are not terminal operators or liner companies but enterprises that require a large amount of imported raw materials and/or fuels. As these enterprises are mainly for domestic consumption rather than for export, they need more favorable conditions to facilitate their connection to the domestic market, such as convenient land transportation system and the availability of the markets near the ports. However, creating these conditions is beyond the compass of ambitious local governments. What they can do only is to reduce the initial investment cost of these enterprises. Therefore, the intervention of the provincial-level government becomes critical.

The Caofeidian port area in the city of Tangshan in Hebei Province is a typical case. To develop its marine economy, the Hebei provincial government fully supported the linking of Tangshan to the Caofeidian Island (18 km away from the coastline and 34 nautical miles away from Tianjin Port) through sea reclamation. Tangshan built a large deepwater port in the island, which is 70 km away from either the city of Tangshan or the original Tangshan Port. It was the only location with a water depth of more than 18 m in Hebei Province. The port was constructed for two reasons: to move Shougang Group to the island and to build Heibei's own access to the sea. Hebei is one of the coastal provinces with the weakest economy. The provincial government regarded Caofeidian as the "No. 1 project in the province" and tried to win over the central government's approval. It aimed to attract heavy chemical industries that were reliant on seaports and raw material importation. In fact, the construction of a port in Caofeidian was not for the expansion of Tangshan but for the creation of a new heavy industrial municipality by establishing large enterprises along the deepwater coast. The newly built township relied much on migrant workers rather than on local residents in Tangshan, as will be discussed later.

Although this kind of intervention occurred in many provinces, it was highly risky though. A recent report reveals that in the case of Caifeidian, Capital Iron and Steel already lost about RMB 10 billion in four years, which was reportedly a fractional part of the total debt of the entire new port city[1]. Another report indicates the amount of debts and loans were so high that the city government of Tangshan needed to pay RMB 10 million a day as loan interest[2]. The major risk was from the irresponsibility and unreliability of the investors at the various

---

1   As reported by Zhang Guodong from First Finance Shanghai on May 28, 2013. http://www.yicai.com/news/2013/05/2736280.html.

2   Wang, Qi: '唐山曹妃甸巨额债务每日利息超千万'   http://money.cnfol.com/130628/160,1538,15429663,00.shtml.

levels of governments. Association of bank loans with the future returns of the projects in the long term, including both port and all near-port industries, were no longer needed. The downturn of the global economy and the slowing down of the Chinese economy exhibited this problem of unreliability.

8.4   Reinterpreting the Port-City Spatial Relationship

Ports (excluding Shenzhen) that have obtained the status as a hub container port in the competition share a common feature. They all experienced a spatial leap forward of dozens of kilometers. This feature was already pointed out several years ago (Wang, 2005). This section attempts to reinterpret this kind of spatial leap forward. Although Dayaowan port area in Dalian, Binhai New Area in Tianjin, Yangshang port area and Luchao New City in Shanghai, Qianwan New Area in Qingdao, Nansha port area in Guangzhou, Haicang New Area in Xiamen, and Beilun port area in Ningbo moved from the original urban areas to deepwater sea shorelines, they leaped forward at different stages of their corresponding city's economy, resulting in different consequences and problems.

The ports with the worst water conditions were the earliest to leap forward. Tianjin New Port in the city of Tanggu (50 km away from Tianjin) and Beilun port area in Ningbo (30 km away from Ningbo) were two new port areas that leaped forward from their original locations in 1980s. They were positioned as ports for handling large bulk cargos (e.g., ores and coal) before the economic reform and opening up. After China entered the period of economic marketization in 1980, some development zones were built before the large-scale container terminal construction was carried out. Therefore, many deepwater coastlines were occupied by enterprises that did not need deep waters much or at all. For example, some food producers, which had their own user's terminals in the Beilun port area in Ningbo, occupied the well-endowed deepwater seashore. A similar situation occurred in Shekou in Shenzhen. The large-scale container terminals could only be built further away. Conversely, in Shanghai and Qingdao, a large number of container shipping-related industries clustered in the old port areas, which could no longer meet the demands of large container ships. Therefore, a large-scale "leap-forward" of port areas occurred. For example, the Qianwan Port district for ocean-going vessels was located 70 km away from the old port and its related industries in Qingdao, and the new port in Yangshan was some 100 km away from the Zhanghuabang port area in suburban Shanghai. At this stage, the relocation for deepwater draft benefited the key parties or stakeholders in the global supply chains – the major shipping lines for cross-continental shipping – as their unit shipping costs were greatly reduced by the deployment of the largest vessels. Conversely, many Chinese export-oriented manufacturers or shippers suffered from the large amount of extra land transportation cost.

For the port city *per se*, the potential effects of the new urban areas surrounding the port are only starting to emerge. These effects are as follows:

1. Construction of large-scale land transport infrastructure, including highway or railway networks of huge capacity;
2. Construction of a system for commuting to the previous city centers and construction of residential areas in the new city;
3. Construction of social and cultural infrastructure required by the new city or town, such as schools, hospitals, and water and electricity supply systems; and
4. Dependence of their development on external factors.

This construction is different from that of traditional new cities or towns because the construction of these new port-surrounding urban areas is not for the people but for the port area itself, which will bring enormous economic benefits. Therefore, the first aspect is the first priority of the local governments, whereas the third aspect is the least. This construction reflects the essence of China's economic globalization: the newly built urban areas surrounding the port areas are no longer traditional cities or towns but new workshops in the global production chain. Furthermore, we also noticed that the workers were not local residents but new immigrants from nearby cities or towns or even further away. Therefore, the new urban areas can benefit from many aspects:

- New port areas can cause the GDP growth desired by the local government during the construction period.
- New port areas can attract migrants from inland places (including those that are losing employment opportunities, e.g., Liaoning Province whose people migrated to the new Bayuquan port area in the city of Yingkou) to coastal areas, as new immigrants' living standard or income will be improved compared with what they used to have in less-developed inland provinces.
- Enterprises (e.g., shipping companies and terminal operators) engaged in the global supply chain can reap the benefits, provided that international or coastal trade keeps growing.
- Port authorities can benefit from the waterfront redevelopment in the old port areas, as they are the nominal landlords of the valuable waterfronts.
- As large-scale producers involved in heavy industries are far away from the urban areas, the environment of the previous densely populated urban areas can be improved.

However, many other effects of these new port areas have not yet been recognized. One of the effects is that new port areas have occupied large areas of deepwater coastlines and lands, including those that should have been reserved for tourism development and environmental protection. Another overlooked effect is the difference between the marine transport network and the land transport network in terms of connectivity. Data analysis carried out by my previous research shows that the average daily shipping capacity of container ships in China's coastal ports reached more than 100,000 TEU in 2007. Supposing that each container held

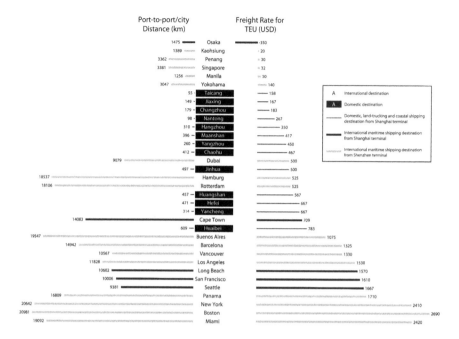

Figure 8.5   Shipping Distances and Freight Rates for Sending a Container from Shanghai or Shenzhen Port to Selective Global and Domestic Inland Destinations (as June 2013)

Source: compiled by author. Freight rates and maritime shipping data are from JCTrans.com and the highway routes from a web-based search engine for highway route and mileage in China: www.jdcsww.com/tools/other/selmile.asp

9 tons of cargos, then 900,000 tons of goods "made in China" per day or 36 million tons per year would be shipped to the world. However, connectivity between these port cities and inland provinces in China was weak. Railways are the weakest one. Most important ports such as Shanghai port and Yantian port in Shenzhen have never reached two percent of their total container throughput handled by rail in the past three decades since the containers were introduced in China.

The land-side transportation fees were even more expensive. Tables 8.4 and Figures 8.5 and 8.6 show the 2013 freight rate of each container (TEU or 20-foot-equivalent unit) from Shenzhen or Shanghai to inland cities is almost the same as that from Shenzhen or Shanghai to some major European ports. The price of shipping each TEU from Chinese coastal areas to most countries in Southeast and Northeast Asia is much lower than that from Chinese coastal areas to cities in the middle part of China via railway. For instance, the rail freight for sending a TEU from Chengdu to Guangzhou by train (USD 883) is more expensive than shipping it from Guangzhou to Rotterdam by ship (USD 600). Highway transport

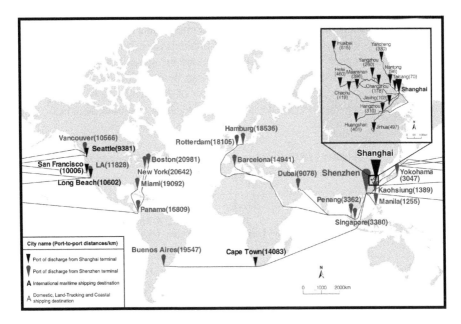

Figure 8.6   Map of Selective Destinations of Shanghai and Shenzhen Ports Listed in Figure 8.5

Note: the number in bracket is the shipping distance in km from Shanghai or Shenzhen port

(Figure 8.5) cost is as high as that of rail transport, and 98 percent or more containers shipped to or from inland in China are by highway trucking. Surprisingly, shipping charge from many Chinese ports to Japanese ports is negative. This occurrence has been going on since 2003. Some analysts[3] assert that the oversupply was done on purpose by some state-owned Chinese shipping lines, with the objective of forcing out the Japanese shipping lines in the market. This strange phenomenon has not disappeared in 10 years but is spreading out to other nearby markets such as Sino–Korea and Southeast Asian routes.

The comparison demonstrates an important transformation in China: the development of coastal ports in recent years has not only turned China into the world's factory but also made the surrounding Asian countries (coastal cities) as connected to China as its inland cities, which bears a strong resemblance to the situation 1000 years ago. Understanding this point is important for the future of China's economic and political geography. Tables 8.4 and Figures 8.5 and 8.6 demonstrate a divergent

3 See ZHOU, Anning (2007) 'Study on the China's Container Shipping Route between China and Japan "Negative Freight" Price Competition', *East China Economic Management*, Vol 21 No.6. 114–17; and TIAN, Zheng and LENG, Lili (2008) "Analysis of 'Negative Freight' on China-Japan Route" *China Market*, Vol. 10, 48–51.

trend between cost-distance and time-distance in multimodal container transport. To
the coastal port cities in China, many Asian destinations are in both low-cost distance
and short-time distance, many inland destinations are in high-cost distance and short-
time distance, and European and North American destinations are in middle-cost
distance and long-time distance. The implications of this trend will be evaluated in
the final chapter in relation to the development of regional free-trade schemes.

**Table 8.4   Container (TEU) Railway Freight Rates from Chengdu to Major
Coastal Port Cities in China**

Origin	Destination	Rate (USD)	Service and Insurance charge	Total charge	Charge (RMB)	Distance (km)	Time (day)
Chengdu	Dalian	916	179	1,095	6,570		12
Chengdu	Qingdao	667	179	846	5,077		7
Chengdu	Shanghai	686	179	866	5,194		7
Chengdu	Lianyungang	588	179	767	4,603		7
Chengdu	Ningbo	863	179	1,042	6,251		7
Chengdu	Guangzhou	883	179	1,063	6,375		7
Chengdu	Shenzhen	915	179	1,094	6,566		12
Chengdu	Xiamen	929	179	1,109	6,651		10

Note: The rates include insurance and service charges. They were implemented in December
2012

Source: Chengdu Logistics Web, accessed from http://028-56.com/tielu/jzxy/2292.html

8.5   Enclaves Created by Port-Logistics-Trade Spatial Integration

The development of China's coastal cities is closely related to various convenient
and special policies on logistics in the international community. June of 1990
saw the establishment of the first bonded zone in China (i.e. Waigaoqiao in
Shanghai). China now has 15 bonded zones (i.e. Waigaoqiao, Tianjin Port, Dalian,
Zhangjiagang, Shatoujiao in Shenzhen, Futian in Shenzhen, Fuzhou, Haikou,
Xiangyu in Xiamen, Guangzhou, Qingdao, Ningbo, Shantou, and Yantian in
Shenzhen and Zhuhai) and an economic development zone (i.e. Yangpu, Hainan)
that enjoy privileged bonded zone policies. Bonded zones are special areas within
a country's territory that facilitates the coming and going of international cargos.
In the 1990s, bonded zones mainly served as warehouses for pre-tax imported
goods to maintain better coherence in the shipping and timing of internationally
traded goods before and after passing Customs.

As hub container ports become increasingly developed, leading port cities in the first round of competition began to usher in another stage of development – setting up relatively mature port bonded logistics parks (or bonded port zone). Since the establishment of the first national bonded port zone in Yangshan port area in Shanghai was approved by the General Administration of Customs in China, the establishments of Dayaowan bonded port zone in Dalian, the Dongjiang bonded port zone in Tianjin, and the Yangpu bonded port zone in Hainan were successively approved. Bonded port zones are the most open and free economic zones in China compared with previous development zones, bonded zones, export processing zones, and bonded logistics parks. The key feature of these zones is that they are located in special places for international logistics and trade that are "inside the territory but outside Customs". In these places, port operations, logistics, and simple processing activities are integrated. In this relatively closed environment, goods are no longer fully supervised by Customs as in bonded zones or bonded logistics parks. Rather, the goods are like inside an outside-Customs "enclave" where the coming and going of people are not supervised and the flows of goods are regarded as entering or exiting the national boundary. Although this kind of enclave is not a 100 percent free trade zone[4], three very important functions exist:

1. No Customs clearance is needed for international water-to-water transshipment in these zones that incorporate some port terminals or specific port areas into the bonded port zones. This way, the terminals or port areas become real places that are outside the Customs and can serve as places where international transshipment and simple processing of goods (e.g., repackaging, unpacking, and transshipment) take place.
2. Foreign goods entering these zones are not regarded as imported. Therefore, goods from other countries put into the same containers with other goods for export from China and then shipped to the international market can enter this zone without going through various Customs clearance procedures and tariff collection.
3. Inland goods arriving at these zones are regarded as having been exported. In China, some goods are exported first and then imported because of special reasons such as trade agreements. These goods can enter these zones first and then make a "U-shaped" return without actually going to other countries.

With these three functions, the port bonded logistics park is similar to a mini Hong Kong that can function as a port, a logistics center, a processing site, and a place for exhibition. Businesses such as port operation, transshipment, international distribution, international sourcing, re-export, export processing, and exhibition can also be developed in this zone. Therefore, for hub container ports with good facilities and internal coordination, establishing this kind of zones will be an advantage.

---

4   See the relevant discussion in Chapter 6 of this book for reference.

The significance of the gradual evolution from the original development zones, export processing zones, bonded zones, and bonded logistics parks (type B)[5] should be thoroughly discussed. The evolution of Tianjin Port from a development zone to a bonded logistics park (Wang and Olivier 2006) was discussed in another study, in which this kind of "economic enclave" as an international articulation space was pointed out. This study refers to the development process as the transformation from building development zones to attract foreign investment and constructing "world factories" in coastal cities to building a hub for international trade and logistics. Bonded port zones are not meant to attract the manufacturing industry but rather international logistics activities. They are mainly about consolidation or deconsolidation of LCL (less-than-container load) in the logistic distribution places, and shipping goods from several countries to specific countries or shipping goods from one country to multiple countries. The role of zones is to operating several international value-added warehouses outside the Customs in these coastal cities. Undoubtedly, these zones are a product of China's active involvement in economic globalization and are the result of the re-regionalization driven by globalization.

8.6   Third Stage of the Coast-Centric Economic Development

The first two stages are not only the result of China's reform and opening-up and the influence of the globalized economy but are also the result of the process of the middle stage of China's industrialization, or what is called the Xiaokang process. However, since the 11th Five-year Plan in 2006, coastal provinces and cities in China have entered the middle-income stage of developing countries. The average annual GNP has reached USD 3,000 to USD 7,000 dollars. One of the important characteristics of this stage is that it begins to enter consumer society; people begin to have time and money. Having nearly 300 million people travel during the golden week in May is a vivid depiction of this stage. In Shenzhen and Guangzhou, thousands of new cars have been registered, which is another reflection of the booming consumer life. As people's income increases and their consumption increases, we also observe the following important processes that occur at the same time:

1. The long process of the RMB appreciating against the US dollar begins. A financial economist from Harvard University considers that this process will not complement China until it becomes the number one economic system in the world, which will be around 2025. According to his

---

5   Bonded logistics parks have two types: A and B. Type A is set up by one enterprise as the legal person and provides bonded logistical services under the supervision of Customs. Type B is formulated by the spatial concentration of several enterprises' bonded logistical services supervised by Customs.

calculations, in real terms, the current exchange rate between the RMB and the dollar is about 4:1. In 2025, the rate should be 1:1. As long as China continues to open up its economy and integrate with the world's economy, RMB appreciation is bound to occur, a factor that also reflects the improvement of real productivity, economic strength, and real living standards of the Chinese people. Although no one can predict how the RMB will appreciate, it is bound to happen, which also means that when buying foreign goods, the RMB is worth more.

2. As more and more middle-end products are produced in coastal cities in China, such as Motorola cell phones produced in Tianjin and Canon cameras produced in Zhuhai, many export-oriented enterprises gradually attach equal importance to foreign and domestic markets with coastal cities and inland capital cities as the major consumption targets. For coastal cities, raw materials and semi-finished products are increasingly complementary. Therefore, trade exchanges between coastal cities become increasingly frequent. As indicated by the abovementioned characteristics, Guangzhou, Tianjin, and other major port cities have seen a higher overall domestic trade growth rate than overall foreign trade for several consecutive years.

3. The increasing number of coastal cities has their own deepwater terminals and is beginning to be served by international liner companies that directly connect them to the whole world. A similar situation occurs in airports that did not originally have international routes. Many second- and third-tier cities (e.g., Hangzhou, Xiamen, and Ningbo) have gradually opened their international direct service. We can expect China's international transport market to be more open, such as opening navigation rights, drafting "open-skies" code-sharing policies, and others. These kinds of policies enable cities with deepwater ports and airports to benefit from their accessibility and connectivity to the international market.

4. Finally, the increase in consumption power is accompanied by the freedom of investment and improvement in tourism and services. Trans-provincial tourism, real estate investment and its corresponding credit business, national and even international credit card circulation, and "free travel" across the national borders can all be seen as an increase in capital and people flow. This change shows that every city will discover that more and more of its consumption and investment percentage comes from other places. Some cities witnessed this trend as early as in the beginning of the reform and opening-up, such as Shenzhen. Currently, this trend is spreading across the nation; it is most obvious in cities with international ports.

Therefore, the four processes can be regarded as the prerequisites for the further development of a coastal-centric economy. Unless the central government adopts policies to change urbanization currently driven by market power and international development of the economy, the coastal cities' leading positions will not change, but they will become even stronger. An important reason is that the improvement rate of inland transportation lags behind that of the international exchange channels of coastal cities. Alternatively, if the inland transport networks, specifically the rail freight services, can be significantly improved, and if the regional integration between China and the rest of the Asian countries slows down, China will have a more regionally balanced growth.

Based on the above analysis, the characteristics of the third stage of the coastal-centric economic development can be summarized as follows:

1. Like European countries, individual coastal provinces and cities in China have been becoming increasingly independent in their communication with the international community, reflecting the global trend of "denationalization". To attain independence, departments directly under the central government, such as the local Customs, can fully cooperate with local governments to set up many channels for international exchanges. Ports and airports can have their own direct international routes without seeking help from others. This strategy enables local state-owned enterprises and private enterprises to build their own global networks to directly carry out various economic and even cultural and technological exchanges with foreign countries. "Denationalization" does not mean that the country is no longer useful. Rather, it means that local places can achieve direct foreign exchanges with the permission of the country. As a result, a growing number of cities are joining the "transnational city system"[6]. Whether they can and how they will integrate themselves into this system will be the key to the development prospect of many Chinese coastal cities and even provinces.

2. Simultaneously, the volume of goods and people exchanges between these cities tends to be great all the time compared with most inland cities, except the provincial capitals.

3. These cities will form their own regional core metropolises and relevant mega city clusters, resulting in the division of labor and coordination within the region, as what already happened in the Yangtze River Delta and the Pearl River Delta. This division of labor is partly caused by market power, historical and cultural accumulation, and government power, or "administrative capital", which varies along the administrative hierarchical levels. From the market point of view, several large cities in the Pearl River Delta have shown clear division of labor: Hong Kong is the regional financial and international trade and logistics center; Guangzhou is the regional hub of domestic trade and logistics and the regional administrative center; Shenzhen has duel characteristics, making it an individual city that leads the national innovation economy as well as the center that provides consumption service and international transportation service to the east of the Pearl River Delta (mainly Hong Kong, Dongguan, Shenzhen, and Huizhou).

---

6   American scholar Saskia Sassen (2000) reveals the existence and operation of such a system in detail in his book *Cities in a World Economy*, second edition.

Macau is the city with the clearest positioning: it is the largest place for gambling and leisure in Asia. Aside from these four cities, several cities dominated by the processing industry have clear regional economic roles to play within the Pearl River Delta, such as Dongguan and Foshan (including Shunde). Regarding cities with historical significance, except Guangzhou, the potential of Zhongshan and Xinhui has not been fully tapped. Zhuhai is closest to the core of the Pearl River Delta and is a "different" city. It has the vaguest role to play. As the central government continues to promote regional development with planning and priority (i.e., the Pearl River Delta, Yangtze River Delta, Bohai-rime, Northeast of China, economic zone on the west side of the Fujian Straits, and economic zone in Beibuwan), external investments and people will continue to flow to the coastal cities with growth potential, whereas local capital and enterprises will gradually set up factory branches or even relocate to cities in the middle and west of China to explore the inland market. The two trends constitute a gradient geography and the tier transfer of national development.

8.7   Summary and Conclusion

This chapter introduced the changes in coastal Chinese cities under the influence of globalization in three layers. Geographically, in the past 30 years, the rapid development brought about by globalization is largely different from the development of the coastal cities of ruling nations (e.g. Liverpool in England) in the period of colonization after the industrial revolution and the coastal cities of the colonies (e.g. cities in Asia, Africa, and South America). First, international market power needs China's cheap labor more than natural resources. Therefore, the competition between China's coastal cities has been changing China's entire coastal area into a global market-driven economic zone with multi-gateways, multi-hubs, and multi-port clusters. The long-distance transport capacity connecting these cities is not land oriented but sea oriented. Railway capacity technically lags behind in transporting containers because multi-modal container transport systems with marine shipping and highway transportation serve the global supply chain rather than the inland market in China. Physically and geographically, the objectives of transnational corporations interested in manufacturing in this country for the global market have accelerated the growth of large port cities. Moreover, with the determination of the local government, the port districts were pushed to leap forward to new areas. Even from the perspective of the city, industries relying on ports tend to be separated from the old port area in terms of function and space. The integration of port, logistics space, and trade space has also experienced a "window effect" from the micro-level geophysical point of view, which reflects an "enclave" of meta-geographical phenomenon with a common meaning. These phenomena are a reflection of global production chains and the development of a global production network (Dickens et al. 2001).

These three layers of coastal city development as a whole do not only show a geographic trend but also reflect the connection of coastal Chinese cities to the global economic network in a special urban spatial structure. In this connection process, the government at all levels tries every possible method to encourage many foreign and domestic investments and enterprises to stay in the areas near the port. These methods include giving various incentives, subsidies, and infrastructure such as port building and setting up international logistics parks. These enterprises change their investment location from a large one (i.e. cities/ market areas) to a small one, considering their connectivity and the convenience of competing in global and/or domestic markets. The governments do the same. They prefer small enterprises to large ones. Any exception occurring in the "strong–strong" and "weak–weak" mutual selection process is the result of the involvement of a higher-level government (e.g. the cases of Qinzhou in Guangxi, Caofeidian of Tangshan in Hebei, and Yangpu in Hainan). The reason why large multinationals choose major port cities as the best investment destination is no surprise: compared with small- and medium-sized cities, major port cities have better infrastructure. Moreover, even in the most expensive port cities in China, such as Shanghai, Shenzhen, Tianjin, and Qingdao, factors of production such as land, labor, talent, and communications for the same quality are still cheaper than those in the United States and in developed countries in Europe.

The global supply chain established in coastal Chinese ports by multinationals (including Chinese multinationals such as the Baogang Group) is the core of the abovementioned process. The "administrative enclaves" of integrated international logistics, trade, and port areas launched by cities such as Shanghai give more advantages to cities that are building the trade pathways. As the initial form of a free trade zone, these "administrative enclaves" will not only become the relay platform of multinational outsourcing but will also reflect that China can integrate itself into a global economic network in various ways. These types of "administrative enclaves" are expected to appear in more cities in China to further reduce the trade barrier. They will not only appear in marine ports but also in airports. If such is the case, more Chinese cities will have lower trade exchanges with the global economy. However, the core of this kind of close trade exchange is that some cities separate the special zones to gradually turn the international trade border to the national territory. In other words, as the earth is becoming "flatter and flatter" under the pressure of globalization, time and cost are becoming smaller and smaller. If the world considers China's economy to be increasingly integrated with the global economy, the differences between Chinese cities and between different areas of the same cities with direct exchange with the world will become greater. This finding is opposite to the European Union experience: Europe is trying to reduce as many barriers and gaps as possible between its member countries. Although the move in China cannot yet be determined as right or wrong at this time, this trend does exist. When the corresponding municipal governments work for the approval of the central government to build these administrative enclaves as a means to compete, more analyses on the influence of this new kind of free trade zones-to-be must be conducted.

This chapter points out another change in time and space caused by this trend: the transport cost distance between coastal Chinese areas and Asian cities with connectivity and other Asian countries with strong economies is closer than the distance between coastal Chinese areas and inland cities in middle and Western China. While the functions of "administrative enclaves" improve, the trade cost between coastal Chinese areas and other Asian cities with high connectivity[7] greatly decreases, as the costs to enter the national territory and customs account for a large percentage of the trade cost (Anderson and Wincoop 2004). We boldly estimate that coastal Chinese areas will then have more opportunities to cooperate with these countries with interfaces like "miniature versions of Hong Kong". China must exert effort to balance economic development by reducing the transportation rate between coastal areas and inland cities. The major airport hubs such as Beijing and Guangzhou were able to gradually obtain opportunities to build airport-based free-trade zones. The transportation capacity gap between coastal areas and inland cities is expected to occur in cities with or without major international airports.

In a word, Chinese port cities become more and more open through various means; the speed of openness seems to be faster than the decreasing rate of logistics costs between coastal areas and inland cities. The theory that the globalized economy will bring benefits to cities located in the network links seems true. However, people have neglected the large gap between different areas of the same cities located in the links. These gaps are important materials for the studying of the impact of globalization because they are the traits of the babies produced by globalization in China – China's new port cities. Studying these gaps indicates that the world is becoming smaller and flatter while these cities are becoming larger, but the development within large developing countries like China is becoming more uneven. To reverse this trend of unevenness in the country and to integrate the containerized China with the under-containerized one, huge efforts must be made to improve the national transport networks for both cargo and passengers. These efforts include constructing high-speed rail systems for passengers, which may provide room for conventional rail system to handle more containers, and obtaining higher intermodal handling capacity that integrates railways with highways and inland waterways to reduce total city-to-city cost-distances for cargo movements between the coast and the inland. In this area of improvement, railway system seems to be more difficult unless a substantial change in its governance is made (see Wang et.al. 2012; World Bank 2010).

---

7 According to Spulber (2007), the international trade cost has four parts: (1) transaction cost, (2) transportation fee, (3) cost of entering customs, including tariff and non-tariff cost, and (4) cost of time.

Chapter 9
Conclusion and Foresight

9.1 Three Characteristics of the Interactive Development of China's Port Cities

The preceding chapters have discussed the interplay between ports and cities in China, from single to multiple port-city relationships and from individual to cluster port-city perspectives in several core development regions. The comparison method was employed during the discussion, comparing the similarities and the differences between different cities in China as well as the holistic development between domestic port cities and foreign ones. In comparison with advanced countries and in addition to previous theories on port development such as the Anyport model by Bird, we observe many distinct features of Chinese cases. Prior to elaborating the prospect of future development for China's port cities, a conclusion on the features of specific patterns or processes observed is stated here.

First, though Chinese port cities are simply duplicating the evolutionary process of other cities in their stages of development, port cities in China have undeniably assimilated unprecedented rapid development processes against the backdrop of the country's enormous urbanization and development since 1978 and its sudden entry into globalization after decades of having a closed economy. Unlike the urbanization in many developed economies and the evolution of their port cities, the process in China occurred within a short period of time. For example, the total port container throughput of the Ningbo Port witnessed a 30 percent growth rate in the past decade, while the number of real residents in the new port city of Shenzhen increased from 100,000 to 10 million in just 30 years. Besides, the transition of many Chinese ports from a conventional port to a modern containerized port surpasses those from other countries in terms of speed and scale of expansion, that is, the distance covered from the river mouth to some deep-water coastlines. Shanghai, for one, established the Yangshan port district on an island 100 kilometers away from its original port and beyond its jurisdiction, which has not been performed anywhere previously. **Spatial expansion under the pressure of increasing demand from the global market within a short period of time undoubtedly became the main characteristic of the development of port cities in China.**

Port cities formed within a short time period and at a larger space reflect the process of urbanization in China under the context of globalization. This book aims to analyze the diverse relationship among the features of urbanization. Generally speaking, we observe that Chinese port cities have upgraded by large-scale expansion that normally requires a long period of gradual evolution for smaller cities. For the past 30 years, large coastal cities and urban clusters had followed this pattern that is not only indicative of ports attracting urban residents, but ports as quintessential factor in attracting large industries and as one of the foundations for related industrial or urban development.

Second, the construction of a large-scale port city and its success in trade must have institutional support and adaptation. Institutional adaptation means that spatial expansion and overall increase in activity are contributed by market forces. The past three decades had actually witnessed the formation of a market system and of related economic space with institutional adaptation. During the re-identification of the value of international logistics zones by market forces, several Chinese port cities had expressed their real economic significance. Institutional support is demonstrated by the Central Government paving the way to development by launching new policies such as development zones, bonded areas, international logistics parks, integrated port-trade zones, and free trade port areas. A few of these institutional adjustments had been tested elsewhere, but in China, the selective and the gradual introduction of these spatially bonded zones as a general national strategy exhibited cautiousness in imposing economic reforms and of opening-up the economy. Meanwhile, a variety of flexible implementations of this strategy as a local government intervention, as demonstrated in Chapter 7, may lead to some unpredictable overinvestment risks in infrastructure. These attempts can therefore bring many positive and negative influences, including social and environmental ones. However, under the guiding concept of "Development is of overriding importance", changes had continued to occur since the 1980s. Therefore, **the dynamic policy environment centered on development through practice is the second precondition and feature.**

Third, local Chinese governments do not only support developing ports by establishing development zones and by promoting logistics. They also directly participate in the market competition. In 1994, the State Council bestowed the power of developing and managing ports to local governments. Since then, local governments have become the strongest influence behind port development, and ports became main areas in which cities with similar capabilities compete with each other. Whether a metropolis such as Shanghai or a small city such as Qinzhou in Guangxi Province, the local government mobilizes resources and attempts every possible option to expand its port scale, not to improve the overall capacity of ports but to attract relevant industries (see Charts 4.5 and 4.7 in Chapter 4). From the perspective of local governments and of potential investment corporations, high prices of land in port neighborhood are not good, but enterprises in relevant industries always hope that their location can be close to both the port and the city, which is against the intent of city governments. However, the primary task of governments is to encourage enterprises to settle in their cities. Each city does its best to show their special prowess in performing operations, including basic land preparation and tax exemption for attracting firms, and improving ties with all kinds of ministries and departments at the central and at provincial governments to obtain favorable special policies.

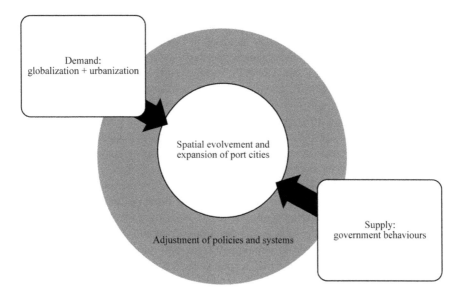

Figure 9.1   The Three Leading Factors Influencing the Development of China's Port Cities

The discounts and other inside stories during the settlement and the establishment of new factories and enterprises near the port area can only be heard through rumors. Even with limited hard evidence, we can observe from another angle that the location and site selections of large port-dependent state-owned enterprises, such as petrochemical firms, are fully based on market rationality. In the process of competing for an industrial complex at a particular scale, port cities in competition do not only rely on seashore or harbor quality, but also the so-called "administrative capital" associated with the administrative ladder that includes the government. When the efforts of the administrative capital are insufficient, the assistance of provincial governments will be solicited. The case of the Guangxi Beibu Gulf Economic Zone discussed in Chapter 8 illustrates this. Therefore, **administrative capital-based behaviors of the local government are the third feature in the formation and development of Chinese port cities.**

The above three characteristics correspond to the three leading factors influencing the development of China's port cities and their spatial expansion. First, with the reformation and opening of the Chinese economy, China has generally transformed from a planned to a market economy with the considerable pressure and strong demand as a result of globalization and of accelerated urbanization. Second, institutional adjustment occurred to withstand the pressure and to meet this demand. Third, governments exert or even induce the necessary pressure and serve as the main players in shaping specific trajectories of port cities (see Figure 9.1).

The players that reflect the orientation of market demands include export-processing enterprises that establish factories in coastal areas, such as Foxconn for Apple iPhone, transnational terminal operators such as Hutchison Whampoa, China Merchants Group, and COSCO Pacific, large state-owned enterprises such as Sinopec, PetroChina, Capital Iron & Steel Group, and Baogang Group Corporation, and a large number of new immigrants and migrant workers who move to coastal cities to start a new life. These new immigrants and migrant workers contribute significantly to the stability and to the development of the society as a whole. Governments have focused for decades on considerable profits and interest brought by the capital and established enterprises of the increasing population, such as the Tianjin Binhai New Area with a permanent population of less than 300,000 contributing 50 percent to the GDP of Tianjin City, from a total of 9 million. In industrial/economic zones, which are close to ports, but far away from the city proper, such as Tianjin Taida Economic Development Zone, Dalian Dayaowan Economic Development Zone, and Ningbo Beilun Economic Development Zone, land is more than a main source of government revenue. Land is used to attract companies from which local governments can gain GDP and taxes. Hence, governments pay more attention to the benefits of globalization.

From another perspective, profitable trade does not only promote the prosperity of ports, but also makes the city more prosperous. Land, the primary source of revenue for local governments, is essential in the process of urbanization. Several studies show that China's city governments tend to actively speed up urbanization to immediately increase revenues (Yi 2009). With the port and the city separated through the course of time or by leaping development (see models in Chapter 4), the city will grow as a result of the separation of value-added spaces for service activities and those for residential to industrial purposes close to the port, as well as provide large low-rent spaces for industries to grow in an ideal location for non-local market. Unlike Western experiences, such as Fos in Marseille, separation and development of a new port district may be conducted rapidly and continuously over a relatively short period of time, and the separation can be distant. One example is the Changxing Island Development Zone of Dalian initiated in 2005. The Changxing Island Development Zone, located about 90 kilometers away from Dalian City, had a similar development to the Dalian Dayaowan Economic Development Zone in the late 1980s, which is 30 kilometers away. The rapid expansion of the port and the industry as a pattern of development inevitably had some problems and risks.

First, enterprises that select locations based on various favorable conditions created by market forces do not necessarily succeed years later. If the enterprises fail or if enterprises entering a new and remote district are limited, a significant portion of land and infrastructure becomes idle. Moreover, industries from ship-building to toy-manufacturing always consume a certain amount of natural resources or even permanently damage the local environment. In "development is of overriding importance", development is only confined to economic development. Without the supervision of media and of other social organizations (i.e. NGOs), governments may overdevelop industries that have high short-term values and considerable environmental costs, rather than developing projects with small exterior cost but with low returns, such as ecological tourism. Furthermore, many large port development zones presently have a common problem of lacking the necessary supporting facilities for living and consumption of new residents. This is because governments and enterprises concentrate on the quality of production space, not on the quality of living spaces. Finally, the location value of ports may be influenced by changes in international or domestic trade routes and portals along with the changing international trade environment. These trade routes and portals may also shift with the varied functions in the globalized economy. A case in point is Dongguan, a city next to Shenzhen Port, known as the "world factory" for electronic products such as mobile phones and computers. From 2009 to 2011, immediately after the world financial crisis, Dongguan's temporal population (i.e. the workers from other provinces) drastically decreased to approximately three million[1] due largely to the closure of a significant number of small and medium-sized firms. As a consequence, the port of Shenzhen also experienced its most serious decline (14 percent) in 2009 on its container throughput, while the port was still rapidly expanding its new port infrastructure in Dachan Bay. If after the crisis, the factories in Dongguan will not return, which is likely the case, the port of Shenzhen, including all terminals under various operators, will face a turning point in handling capacity surplus rather than under-capacity issues for the first time. This situation will present a huge challenge to the city government as well as to port operators in adapting to a series of new institutional changes that will allow coastal land resources to be allocated for other more valuable uses.

9.2   Problems and Prospects in China's Port Cities

9.2.1   From port cities to portals of international terminals

In looking back on the evolution of important port cities around the world, many of them, such as London, New York, Chicago, and Venice, had redirected their focus on their economies from ports to a more advanced structure of the economy. Hong Kong, as another example, in the 1970s to the early 2000s, transformed from a port centered on processing and trade to a structure with a focus on financial and tourism businesses and also on high-end services that include the global supply chain management. In 2012, 40 percent of the trade values on import, export, and re-export were via air transportation and only 28 percent was conducted via maritime transportation. In terms of employment, though a large number of

---

1 According to an internal seminar facilitated by a government official on April 2012, who was confirmed also as an expert in the area of urban planning, the total number of temporal workers is estimated at about 8 million, its highest number during the first half of the 2000s.

personnel have jobs related to logistics, the number of employees solely devoted to port services is diminishing. In 2008, a research conducted on the Hong Kong logistic industry reported that Hong Kong was transforming from a shipping center to a global supply chain management center, therefore combining the advantages of finance, trade, and logistics and forming a comprehensive service platform where real cargo flow of the supply chain do not necessarily go through the Hong Kong port.

With the rapid development, the question remains on whether mainland port cities in China will follow the same trend in the near future. Shanghai, for example, was ranked as the leading global port and international trade city during the first decade of the twenty-first century. Meanwhile, this period also witnessed the fastest rise in land and human resource costs, as well as economic pluralism. The government in Shanghai is aware that the highway transport of 17 million standard containers (TEU) disturbed the normal city life. Currently, Shanghai no longer dwells on how to increase the port throughput, but focuses on how to increase the connecting capacity of transport to divert the port-induced traffic, as well as storage space, away from the core areas of the metropolis. This increase is associated with the allocation of more spaces for other value-added activities and the rapid development of the airport and the railway network. Shanghai can likely achieve this goal as a very strong economic power and administrative capital. The upgrading of Shanghai to a national and international financial center will increase its demands for both ports and airports.

Although Shanghai may not meet any generalization, the essential consideration is on which port can earn benefits and gain high status in becoming eternal portals of international trade along with the evolution of China's future role in the global arena, as well as which cities will continue with their original roles. Chapter 5 proposes a system of external connectivity of the fourth-tier port cities. In my view, this is an increasingly sophisticated system without significant changes in its structure in the future. However, the cities that would have higher opportunities to become regional portals depend on the following improvements or changes:

1. Characteristics and formation of the Asia-Pacific Free Trade Zone and beyond. Since 2010, the Free Trade Agreement between China and ASEAN has been in effect. Japan and Korea are actively communicating with ASEAN in an attempt to form a "10+2" or "10+3" Asian Free Trade Zone. Upon establishment of these free trade zones, trading between China and Asia-Pacific countries will be promoted. For China's port cities with specific international connection with other Asian regions or countries despite major differences, new and more relaxing trading terms may lead to more benefits, especially if they have complementary features for involved cities or regions. New regional portals or gateways may emerge because of better internal relationship with a large potential market and a more holistic preparation of some ports. To a certain extent, this scenario may apply to port cities such as Guangzhou and Yiwu of Zhejing that have growing

trade relations in other parts of the world, such as Africa and Middle East. Nansha Port interestingly plays a role in reestablishing Guangzhou as an entrepôt by forming the new "silk road on the sea" to Africa.

2. Growth of the hinterland market. Several parts of this book have mentioned that China's mainland has still not seen containerization for the entire country. With the coming age and with emphasis on both domestic and foreign markets, containerization in inland cities will be included in the agenda, enabling growth for inland cities and their efficient connections with coastal regions. However, as to which cities have more opportunities to lead the change depends mainly on the railroad business. A drastic change was witnessed in 2013. The Ministry of Railways merged with the Ministry of Transport, and the railway operation was finally incorporated into the China Railways Group. This merging will hopefully speed up the process of the long-awaited containerization in the rail freight sector. In comparison with the improvement in inland water-borne transport, such as that in the Pearl River as discussed in this book and in the Yangtze River, several railway corridors for container transport that started in major port cities seem to be more promising, which may help Qingdao, Shanghai, Tianjin, and Guangzhou to gain more and to become faster than the others.

3. Development of multi-modal transport portals and terminals. This development refers to the establishment of an enterprise or transport union as a pivot point to achieve multi-modal transportation such as sea-railway and sea-river transport. From the end of the last century, the development of multimodal transport and logistics terminals has begun to fundamentally influence the advancement of ports (Notteboom and W. Winkelmans 2001). However, no inland port city has succeeded in achieving advancement because of three main obstacles. First, the railroad lacks the necessary carrying capacity and interest. Second, inland river shipping companies thrive in low-level and low-cost competition and lack the enterprise capability of engaging the industry, particularly in attracting the interest of other transportation and logistics companies interest to invest and to form coalitions. Third, local governments do not provide substantial and effective support.

4. The management, governance, and coordination of the functional zone of the port, logistics, and of the industry. For a port city, one important factor that provides it with an advantage over non-port cities should be its "in-transit efficiency" gain through port and port-related or port-induced transport systems. In-transit efficiency means that when a cargo stops at a port city, it should obtain effective logistics services and more importantly, it should be able to effectively go through all necessary procedures, such as customs inspection, tariff collection, finance, and insurance, in a timely manner. In-transient efficiency demands cooperation among many components that form suitable environment of (global) supply chains, such as the functional zones of international logistics near port, international airport for sample

delivery, business communications, and infrastructures that facilitate all kinds of trade that include e-commerce business. Quality management and governance also have a key role, such as in the standardization of information protocols and of formal, anti-corruption, and anti-smuggling policies. Specific studies indicate that the non-barrier cost generated by these links is a maximum of 40 percent of the total transportation and logistics costs.

9.2.2   From foreign trade resource to waterfront landscape resource

This book mainly aims to explain how China's port cities focus and utilize foreign trade resources while facing the pressure and the demand brought about by globalization and by urbanization. In the future, however, many port cities should consider more importance on the full utilization of their own waterfronts as a landscape resource. Sydney and other cities are good examples that China may learn from. Shanghai, Hong Kong, Guangzhou, Xiamen, and Nanjing have made significant efforts. However, the governments of Shanghai and of Hong Kong do not immediately coordinate their comprehensive plans to improve their waterfront areas. For example, the degree of reclamation at Hong Kong's Victoria Harbor has been too high and many skyscrapers surround the waterfront, leaving a small space for citizen's public use.

The development of waterfront landscape resource may have a different timeline scheme. Shanghai, for example, had a rapid and ambitious plan to redevelop its Pu River by establishing large-scale architecture sites such as the Shanghai World Expo. By contrast, Sydney implemented a long-term and flexible master plan for their redevelopment work, consequently allowing opportunities for architectural diversity instead of all projects conducted at the same time. While the Sydney model of waterfront redevelopment may be applied on Chinese port cities, these cities may experience difficulties because of two reasons. First, owing to the current urban financing structure in the country, city governments tend to rush the increase of its land value for higher taxation, which is the most important source of government revenue. The fast and the large-scale renovation and redevelopment of the old port area with valuable waterfront would be an attractive option. Second, although decades of gradual development can integrate building styles from different ages, providing the city with a unique historical aspect requires not only institutional patience for city governments, but also a cultural and philosophical mentality that is less observed in China today. Ironically, embodying this patience in city planning and in (re)development, to leave a space for future generations, is similar to an essence of Chinese culture indicated by traditional Chinese paintings, which is freehand brush work emphasizing only several corners and leaving most of the painting empty and untouched!

9.2.3   Further evolution of the internal space of port cities

The biggest difference between the internal space of port cities and that of other cities lies in the contrasting land use between a port and its city. With the growth of the port and its city, this difference continuously transfers in space (see Chapter 4), and another important manifestation is that the port's highway transport has a larger negative influence or increased externality or social cost. Shenzhen, Shanghai, and Tianjin are all examples of this manifestation. The high concentration of marine logistics emerged from the "hub and spoke" operation of shipping companies and from the enlargement of regular ships. Eliminating this conflict has become a challenge in city planning and management worldwide. Shanghai-Ningbo and Shenzhen-Hong Kong logistics centers, which are among the top ones in the world, and the Tianjin Binhai New Area, which is filled with large industrial projects, can partly address the problem by building large-capacity routes for railroad use similar to Los Angeles. The key is to avoid many industrial clusters in one area, which is not preferable for current Chinese city governments. Therefore, the most likely consequence is the continuous deterioration of the problem or situations that result in severe accidents or unsuitable road environments.

Another possible, may be better alternative is port regionalization. In the Yangtze River Delta, several medium-sized ports appeared around Shanghai-Ningbo and succeeded in sharing specific advantages, especially in the short-sea market connections to Japan, Korea, Taiwan, and other Southeast Asian countries. Shanghai and Ningbo may not have realized that the growth of these new ports is positive for them because together they comprise an effective and environment-friendly means to ease traffic congestion in existing large pivot ports and their cities.

9.2.4   Sustainable development of port cities

Sustainable development entails economic efficiency, social justice, and environmental protection. The port has always been viewed as a way to provide economic competence for cities and to show that governments are fully aware of its economic benefit. Therefore, more attention should be given to aspects of societal equality and environment. In advanced countries, port-related social problems are mostly about addressing reemployment issues of dock workers caused by ports or terminals that have become obsolete. Currently, China's ports are at a rising stage. The main problem is not the re-employment of workers, but the medical care of migrant workers and of new immigrants, as well as the education of their children and transport needs of workers in new areas of the city, including those working for or in the new port district far away from city center. Many new port areas such as Bayuquan in Yingkou are actually new towns occupied by migrants. The comparison and the balance between the social development of ports and that of the city can be seen as social indicators for the port's sustainable development. Indeed, many social problems of new immigrants living in port or coastal cities are rarely considered as "port-induced" issues because any derived solution does not seem to have a direct link with the port business. For example, the *hukou* or the residential registration system has prevented many people from the countryside to obtain a citizenship in the cities they work in, and this problem is associated with a fundamental nationwide policy on urbanization that cannot be handled individually by a port or a port city. However, as the salary gap between coastal and inland provinces decreases and as the inland economy improves, the unfair treatment to new immigrants in the coastal region, such as in obtaining citizenship may result in the region become less attractive than cities in their home province. The loss of low-paid workers in the coastal regions of China indicates lesser competition in the cities for global production than that in other newly industrialized countries such as Vietnam and Bangladesh. Therefore, social issues eventually affect the demand for ports as a result of changes in the trade pattern in a large geographical scale and in a long period of time.

From the perspective of environmental protection, the port is an important area to be governed. On one hand, the change in natural coastlines, reclamation, and channel excavation due to ports can have many ecological effects. On the other hand, the frequent utilization of fuel of sea-bound vessels, with high sulfur content will result in air pollution that is worse than that caused by cars if these vessels do not change fuels before entering the port city. Water transport is generally better for the environment than highway transport because the former consumes lesser energy due to buoyancy when shipping the same amount of cargo. However, when a number of large-tonnage ships stop at one port, the local concentration of air pollution becomes hazardous to the health of the local people. The previously mentioned highway traffic congestion is also because of large ports. Neither the advantages of water-borne transport nor the pollution caused by bunker fuel in ports and in harbors are unfortunately not addressed properly in most Chinese port cities.

Currently, China's port planning has not considered the social costs of port activities. In city planning and in regional planning, the combination of different means of transportation and of more environment-friendly development was not included in the agenda. Further research is necessary to be done. For example, for international carbon trade, we need to know how much carbon emissions are caused by transportation and logistics in the foreign trade volume through ports. If a city were to change 20 percent of the existing highway transport to railway or inland river transport, how much carbon emissions will be reduced? To obtain these important data, technical support is needed to propose and to implement certain compensation mechanisms. Before the environmental goals can be achieved, some fundamental statistics and analysis should be carried out, as done so in Marseille-Fos, France (Merk and Comtois 2012).

In general, with the increasing export demand for raw materials and fuels that lead to the development of the economy, large processing industries and heavy machinery manufacturers who use raw materials such as petrochemicals, power, iron, and steel, and ship manufacturing industries will continue to move and to gather in coastal areas, especially in areas that are close to ports. Therefore, the sustainable development of port cities does not only matter for cities and ports. In 2009, the large Sinopec-Kuwait refinery was originally planned to be located next to the Nansha Port at the geographical center of the Pearl River Delta. Objections to the decision were eventually heard and the site was moved outside the region to Zhanjiang. The problem is not only the environmental capacity of each city and region, but also the ecological balance of the entire Pearl River Delta and that of the Guangdong Province. This is how sustainable development of port cities is incorporated into the configuration of regional sustainable development, which is common in China and in port cities worldwide against the background of globalization. Are these transport links and facilities conducive to trade? What kind of infrastructural and institutional settings can the port city and its residents achieve a fair share of benefit from the trade handled? Unfortunately, one cannot expect a universally accepted answer to these questions because of the unique qualities of each place economically and politically. However, the exciting part of this geographical research is the exploration of different situations and development paths with a common background and theories to conclude a better understanding of overall causes, and then exploring specific solutions for individual places.

Bibliography

Acemoglu, D. and Robinson, J.A. 2012. *Why Nations Fail: The Origins of Power, Prosperity and Poverty*, Crown Business.

Anderson, J., Van Wincoop, E. 2004. Trade costs, *Journal of Economic Literature*, 42(3): 691–751.

Banga, I. Ed. 1992. *Ports and Their Hinterlands in India*. New Deli: Manohar Publishers and Distributors: 1700–1950.

Berechman, Y. 2007. The social costs of global gateway cities: the case of the port of New York. International Conference on Gateways and Corridors 2007. Vancouver.

Beresford, A.K.C., Gardner, B.M., Pettit, S.J., Naniopoulos, A. and Wooldridge, C.F. 2004. The UNCTAD and WORKPORT models of port development: evolution or revolution? *Maritime Policy & Management*, 31(2): 93–107.

Bichou, K. and Gray, R. 2005 A critical review of conventional terminology for classifying seaports, *Transportation Research Part A*, 39(1): 75–92.

Bird, J. 1963. *The Major Seaports of the United Kingdom*. London: Hutchinson and Co. Ltd.

Bird, J. 1971. *Seaports and Seaport Terminals*. London: Hutchinson and Co. Ltd.

Bird, J. 1983. "Gateways: slow recognition but irresistible rise". *Tijdschrift Voor Economische En Sociale Geografie*, 74(3): 196–202.

Brooks, M.R. 2004. The governance structure of ports. Review of Network Economics: Special Issue on the Industrial Organization of Shipping and Ports, 2(2): 169–84.

Brooks, M.R., Cullinane. K. 2007. *Devolution, Port Governance and Port Performance.* London: Elsevier.

Buck Consultants International (The Netherlands), ProgTrans (Switzerland), VBD European Development Centre for Inland and Coastal Navigation(Germany), via donai (Austria). 2004. Prospects of Inland navigation within the enlarged Europe http://ec.europa.eu/transport/inland/studies/doc/2004_pine_report_report_full.pdf accessed on 2010-03-11.

Burghardt, A.F. 1971. "A hypothesis about gateway city". Annals of the Association of American Geographers, 61(2): 269–85

Button, K. 2007. Distance and competitiveness – emerging continental network barriers and strategic partners. International Conference on Gateways and Corridors 2007. Vancouver.

Charlier, J. 1996. The Benelux Seaport System. Tijdschrift voor Economische en Sociale Geografie, 87(4): 310–21.

Chen, Hang. 2006. Study on Port-City Relationship in Dalian, [Master Dissertation], Dalian: Liaoning Normal University (Chinese).

Cheng Siwei. 2003. From Bonded Zones to Free Trade Zones: Reform and Development of China's Bonded Zones. Beijing: Economic Press (Chinese).

China Academy of Urban Planning and Design. 2006. Study on Urban Spatial Development Strategies in Fuzhou, Manuscript at http://xiazai.dichan.com/show-240948.html, accessed on 2010-03-11.

Christaller, Walter (1933): *Die zentralen Orte in Süddeutschland*. Gustav Fischer, Jena.

Dai, Dunfeng and Wang Shuting. 2009. Lu Dadao, Hu Zhaoliang and Zhou Yixing: Their Theories can Influence China's Economic Layout. Chinese National Geographic (中国国家地理杂志), 10: 190–209 (Chinese).

Debrie, J., Guovenal, E. and Slack, B. 2007. Port devolution revisited: the case of regional ports and the role of lower tier governments. *Journal of Transport Geography*, 15: 455–64.

Dicken, P., Kelly, P.F., Olds, K., Yeung, H.W.-C. 2001. Chains and networks, territories and scales: towards a relational framework for analyzing the global economy. *Global Networks*, 1(2): 89–112.

Dou, Ping. 2006. From Bonded Zone to Free Ports – Study on the Upgrade of Functions of Shanghai Bonded Zones : [Master thesis]. Shanghai: East China Normal University (Chinese).

Ducruet, C., Lee, S.-W. 2006. Frontline Soldiers of Globalization: Port-City Evolution and Regional Competition. *GeoJournal*, 67(2): 107–22.

Fan, Ruseng. 2005. Tianjin Port Trade and the Development of Export-Oriented Economy in Hinterland (1860–1937): [Doctoral thesis]. Shanghai: Fudan University (Chinese).

Fremont, A. 2007. Global maritime networks: The case of Maersk. Journal of Transport Geography, 15(6): 431–42.

Frenken, K., van Oort, F. and Verburg, T., 2007 Related variety, unrelated variety and regional economic growth. Regional Studies, 41: 685–97.

Fujian Provincial Government. 2009. Planning of Highway Network Layout in Economic Zone on the West Side of Fujian Straits. (2009–06–15), accessed on 2010-3-11 at http://www.fjjtghb.cn/Article_Show.asp?ArticleID=352 (Chinese).

Fujita, M. and Mori, T. 2005. The role of ports in the making of major cities: self-agglomeration and hub-effect. *Journal of Development Economics*, 49(1): 93–120.

Fuzhou Municipal Construction Planning Commission. 1999. Mater Urban Planning of Fuzhou City　(1995–2010), http://www.chinasus.org/case/general/20090729/51237-2.shtml, accessed on 2010-3-11 (Chinese).

General Administration of Customs, National Development and Reform Commission, Ministry of Finance, Ministry of Land and Resources, Ministry of Commerce, State Administration of Taxation, State Administration of Industry & Commerce, State Administration of Quality Supervision, Inspection and

Quarantine, State Administration of Foreign Exchange. 2004. Approval Criteria and Procedures to Build Export Processing Enterprises (2004-04-08). http://www.law110.com/law/other/19164.htm, accessed on 2010-03-11 (Chinese).

Gereffi, G., Humphrey, J. and Sturgeon, T. 2005. The governance of global value chains. *Review of International Political Economy*, 12: 78–104.

Gillen, D. and Parsons, G. et al. 2007. Pacific crossroads: Canada's gateways and corridors. International Conference on Gateways and Corridors 2007. Vancouver.

Haikou Customs of the People's Republic of China. 2007. "Questions answered by Customs Director Li Xuelan in the Press Conference on Yangpu Bonded Zone Convened by Haihai Provincial Government", government document, (2007-10-12) (Chinese) http://haikou.customs.gov.cn/publish/portal128/tab9101/module29287/info85490.htm accessed on 2010-03-11.

Hall, P.V. 2007. Global logistics and local dilemmas. International Conference on Gateways and Corridors 2007. Vancouver.

Hall, P.V. and Hesse, M. 2012a. Reconciling cities and flows in geography and regional studies. In: Hall, P.V., Hesse, M. (Eds), Cities, Regions and Flows: 3–20.

Hall, P.V., Hesse, M., 2012b. Cities, flows and scale. Policy responses to the dynamics of integration and disintegration. In: Hall, P.V., Hesse, M. (Eds), Cities, Regions and Flows, pp. 247–59.

Hall, P.V. and Jacobs, W. 2010 Shifting Proximities: The Maritime Ports Sector in an Era of Global Supply Chains, *Regional Studies*, 44:9, 1103–15.

Hall, P.V. and Jacobs, W. 2012 Why are maritime ports (still) urban, and why should policy-makers care?, *Maritime Policy & Management*, 39(2): 189–206.

Hayuth, Y. 1987. Intermodality: Concept and Practice. London: Lloyd's of London Press Ltd.

Hayuth, Y. 2007. "Globalisation and the port-urban interface: conflicts and opportunities". in Wang. J. et al. (Eds). 2007. Ports, Cities, and Global Supply Chains. Aldershot: Ashgate: 141–56.

Hendersen, J., Dicken, P., Hess. M., Coe, N. and Yeung, H.W.-C. 2002. Global production networks and the analysis of economic development. *Review of International Political Economy*, 9: 436–64.

Hesse, M. 2013 Cities and flows: re-asserting a relationship as fundamental as it is delicate, *Journal of Transport Geography*, 29: 33–42.

Hesse, M. 2010. Cities, material flows and the geography of spatial interaction. Urban places in the system of chains. Global Networks, 10(1): 75–91.

Hesse, M. 2008. The City as a Terminal. The Urban Context of Logistics and Freight Distribution. Ashgate, Aldershot.

Hesse, M. and Rodrigue, J.-P., 2006. Global production networks and the role of logistics and transportation. *Growth and Change*, 32 (4): 499–509.

Hilling, D. 1996. *Transport and Developing Countries*. New York: Routledge.

Hilling, D. and Hoyle, B.S. 1984. "Spatial approaches to port development". In Seaport Systems and Spatial Change: technology, industry, and the development strategies. Chichester, Sussex: John Wiley & Sons: 1–19.

Hoyle, B.S. 1968. East African seaports; an application of the concept of "Anyport". Transactions and Papers of the Institute of British Geographers, 44: 163–83.

Hoyle, B.S. 1989. The port – city interface: trends, problems and examples. *Geoforum*, 20: 429–35.

Hoyle, B.S. and Pinder. D.A. (Eds). 1981. Cityport Industrialization and Regional Development. Oxford: Pergamon Press.

Huang, Shengzhang. 1951. The Development of China's Port Cities. Journal of Geographical Science, 18(1): 21–40 (Chinese).

Huo, Jiacai. 2009. A Brief Discussion on Building Port to Develop its Located City. Coastal Enterprises and Science & Technology, 7: 34–5 (Chinese).

Jacobs, W. 2007. Political Economy of Port Competition. Institutional Analysis of Rotterdam, Dubai and Southern California (Nijmegen: Academic Press Europe).

Jacobs, W. and Notteboom, T.E. 2011. An evolutionary perspective on regional port systems: the role of windows of opportunity in shaping seaport competition. *Environment and Planning A*, 43(7): 1674–92.

Jacobs, W., Koster, H. and Hall, P.V. 2011 The location and global network structure of maritime advanced producer services, *Urban Studies*, 48: 2749–69.

Jessop, B. and Sum, N.-L. 2000. An Entrepreneurial City in Action: Hong Kong's Emerging Strategies in and for (Inter-) Urban Competition. Urban Studies, 37(12): 2287–310.

Krugman, P. and Elizondo, R.L. 1996. Trade policy and the Third World metropolis. Journal of Development Economics, 49: 137–50.

Langen, P.W. de (2004) The Performance of Seaport Clusters; a Framework to Analyze Cluster Performance and an Application to the Seaport Clusters of Durban, Rotterdam and the Lower Mississippi, ERIM and TRAIL thesis series.

Li Jialin, Zhu Xiaohua and Zhang Dianfa. 2008. Characteristics of Land Use Expansion of Clustering Port Cities and their External Pattern Evolvement – – Taking Ningbo as an Example, Geographical Research, 27(2): 275–85 (Chinese).

Li, Fei, Shishi, Li and Shanxiang, Hong. 2003. Interpretation of the Law of the People's Republic of China on Ports. Beijing: Law Press (Chinese).

Lin, G.C.S. and Yi, Francine 2011 "Urbanization of Capital or Capitalization on Urban Land? Land Development and Municipal Finance in Urbanizing China". *Urban Geography*, 32(1): 50–79. Columbia, USA: Bellwether.a.

Liu, Hui. 2003. Study on the Economic Performance of Shanghai Development Zone and its Development Model: [Master thesis]. Shanghai: East China University (Chinese).

Liu, Yan. 2004. Study on Mutual Relationship between Bonded Zones and Port Development – Taking the Consociation Development between Waigaoqiao Bonded Zone and Port Area: [Master thesis]. Shanghai: Shanghai Maritime University (Chinese).

Liu, Zhiqiang. 2005. On Port and Industry Clustering: [Master thesis]. Shanghai: Shanghai Maritime University (Chinese).

Mao, Boke. 2005. On Port Space. China Port, 4: 11–13 (Chinese).

Mason, K. 2007. Is the "Gateway" concept useful or relevant to the passenger aviation market? International Conference on Gateways and Corridors 2007. Vancouver.

Ministry of Transport of the People's Republic of China. 2006 "Pan-Pearl River Delta Highway Planning (see Planning Outline of Infrastructure of Highway and Water Transportation for Regional Cooperation in Pan-Pearl River Delta)". Government Document, (2006-03-03) http://www.pprd.org.cn/ziliao/zhengce/guihua/200607/t20060717_9796.htm accessed on 2010-03-11 (Chinese).

Morrison, W.G. (2007). Gateways and Corridors: Ten Messages. International Conference on Gateways and Corridors 2007.

National Development and Reform Commission. 2008. Development Planning of Beibuwan Economic zone in Guangxi (2006–2020). Government Document (2008-02-21) http://www.gxzf.gov.cn/C9/C9/gxsywgh-zxgh/Document%20 Library/gxbuwzjqfzfh.pdf, accessed on 2010-03-11 (Chinese).

National People's Congress. 2003. The Law of the People's Republic of China on Port. http://search.moc.gov.cn:8080/was40/, accessed on 2010-3-11 (Chinese).

Notteboom, T.E. 1997. Concentration and load centre development in the European container port system. Journal of Transport Geography, 5(2): 99–115.

Notteboom, T.E. 2006 Concession Agreements as Port Governance Tools. *Research in Transportation Economics*, 17: 437–55.

Notteboom, T., Ducruet, C. and de Langen, P. (Eds). 2009. Ports in Proximity. Competition and Coordination among Adjacent Seaports. Ashgate, Aldershot.

Notteboom, T.E. and Rodrigue, J.P. 2007. "Re-Assessing Port-Hinterland Relationships in the Context of Global Commodity Chains". In: Wang. J.J., Olivier. D., Notteboom, T.E. and Slack, B. (Eds). *Ports, Cities and Global Supply Chains*. Aldershot: Ashgate: 51–68.

Notteboom, T.E. and Rodrigue, J.-P. 2005. Port regionalization: towards a new phase in port development. *Maritime Policy and Management*, 32(3): 297–313.

Notteboom, T.E. and Winkelmans, W. 2001. Structural changes in logistics: How do port authorities face the challenge. *Maritime Policy and Management*, 28: 71–89.

Pain, K. 2007. Global cities, gateways and corridors: hierarchies, roles and functions. International Conference on Gateways and Corridors 2007. Vancouver.

Pallis, A.A. and Syriopoulos, T. 2007. Port governance models: Financial evaluation of Greek port restructuring. *Transport Policy*, 14(3): 232–46.

Pettit, S.J. and Beresford, A.K.C. 2009. Port development: from gateways to logistics hubs, *Maritime Policy & Management*, 36(3), 253–67.

Pinder, D. and Slack, B. (Eds). 2004. *Shipping and Ports in the Twenty-first Century: Globalisation, Technological Change and the Environment*. London: Routledge.

Provincial Department of Transport in Fujian. 2006. Special Planning of Comprehensive Transport System in the Economic Zone on the West Side of Fujian Straits During the "Eleventh Five-year Period". Accessed on 2010-3-11, at http://www.drcnet.com.cn/DRCNet.Common.Web/docview.aspx?version=Integrated&docid=2102856&leafid=14009&Chnid=3579 (Chinese).

Rimmer, P.J. 1967. The search for spatial regularities in the development of Australian seaports 1861 – 1961/2. *Geografiska Annaler Series B Human Geography*, 49(1): 42–54.

Robinson, R. 1976. Modelling the Port as an Operational System: A Perspective for Research. *Economic Geography*, 52(1): 71–86.

Robinson, R. 1985. Industrial strategies and port development in developing countries: the Asian case. Tijdschrift voor Economische en Sociale Geografie, 76: 133–43.

Rodrigue, J.-P. 2007. Gateways, corridors and global freight distribution: the pacific and the North American maritime/land interface. International Conference on Gateways and Corridors 2007. Vancouver.

Rodrigue, J.-P. and Notteboom, T. 2010. "Comparative North American and European gateway loggistics: the regionalism of freight distribution". Journal of Transport Geography, 18: 497–507.

Rodrigue, J.-P., Comtois, C. and Slack. B. 2009. The Geography of Transport Systems. New York: Routledge.

Rostow, W. 1960. The Stages of Economic Growth: A Non-Communist Manifesto. Cambridge: Cambridge University Press.

Sassen, S. 2000. Cities in a World Economy. Newbury Park: Pine Forge Press.

Shi, Youfu and Jingning, Hui. 1996. The Influence of Port on Urban Economic Development. China Water Transport, 5: 9–11 (Chinese).

Shi, Youfu and Jian, Zhou. 2005. The Positioning and Development Prospect of Top Eight Chinese Container Ports in International Competition. China Port, 6: 34–7 (Chinese).

Shi, Youfu. 2003. Port-City Relationship and the Reform of Port System. China Port, 1: 14–16 (Chinese).

Slack, B. 1993. The impacts of deregulation and the US-Canada free trade agreement on Canadian transportation modes. *Journal of Transport Geography*, 1(3): 150–55.

Song, Bingliang. 2001. Estimation of Overall Shanghai Port Economic Contribution. Journal of Shanghai Maritime University, 22(4): 11–14 (Chinese).

Song, D.-W. 2002. Regional Container Port Competition and Co-operation: The Case of Hong Kong and South China. *Journal of Transport Geography*, 10: 99–110.

Spulber, D.F. 2007. Global Competitive Strategy. Cambridge: Cambridge University Press.

Swyngedouw, E. 2004. Globalisation or 'Glocalisation'? Networks, Territories and Rescaling. Cambridge Review of International Affairs, 17(1): 25–48.

Taaffe, E.J., Morrill, R.L. and Gould, P.R. 1963. Transport expansion in underdeveloped countries: a comparative analysis. *Geographical Review*, 53: 503–29.

The General Office of the State Council of the People's Republic of China. 2001 "Opinions on Deepening System Reform of Department Directly under the

Central Government and Dual Leadership of Port Management", government document, (2001-11-23) (Chinese) http://www.yfzs.gov.cn/gb/info/LawData/gjf2001q/gwyfg/2003-08/04/1506529460.html accessed on 2010-03-11.

The State Council of the People's Republic of China. 1985. Interim Provisions of the State Council of the People's Republic of China on Preferences for the Construction of Ports and Piers with Chinese and Foreign Joint Investment. http://www.gzas-l-tax.gov.cn/taxlaw/show.asp?id=1392 accessed on 2010-03-11 (Chinese).

Tongzon, J.L. (2007). The role of port performance in gateway logistics. International Conference on Gateways and Corridors 2007. Vancouver.

Tretheway, M.W. and Andriulaitis, R. 2007. Gateway & corridor performance: what is important? International Conference on Gateways and Corridors 2007. Vancouver.

United Nations Conference on Trade and Development (UNCTAD). 1992. Port Marketing and the Challenge of the Third Generation Port. Geneva: UNCTAD Press.

Van Klink, H. and Arjen, et al. 1998. "Gateways and intermodalism". Journal of Transport Geography, 6(1): 1–9.

Van Klink, H. 2002. The Kempun Nexus. The spatial-economic development of Antwerp and Rotterdam, in Loyen, R., Buyst, D. and Devos, G. (Eds) *Struggling for Leadership: Antwer-Rotterdam Port Competition 1870–2000,* Heidelberg: Physica: 141–60.

Wang, Chengjin and Fengjun, Jin. 2006. Study on the Organization Network of Chinese Marine Container Transportation. Geographical Science, 26(4): 392–401 (Chinese).

Wang, J.J. and Chen. C.M. 2010. From a hub port city to a global supply chain management center: a case study of Hong Kong. Journal of Transport Geography, 18: 104–15.

Wang, James and Ng, Adolf (2010) "The Geographical Connecedness of Chinese Ports with Foreland Markets: A New Trend?", Tijdschrift voor Economische en Sociale Geografie, Vol. 102: 188–204.

Wang, James J., Rong, C., Xu, J. and Or, S.W.O. (2012) "The funding of hierarchical railway development in China", Research in Transportation Economics, Vol. 35: 26–33.

Wang. J.J. 1998. A container load center with a developing hinterland: a case study of Hong Kong. *Journal of Transport Geography*, 6: 187–201.

Wang. J.J. 2011. "Entrepreneurial Region and Gateway-making in China: A case Study of Guangxi", a chapter in Integrating Seaports and Trade Corridors, edited by Hall, P., McCalla, B., Comtoice, C. and Slack, B. Farnham: Ashgate: 247–60.

Wang. J.J., Ng. A.K.-Y. and Olivier. D. 2004. Port governance in China: a review of policies in an era of internationalizing port management practices. *Transport Policy*, 11: 237–50.

Wang. J.J. and Olivier. D. 2003. La gouvernance des ports et la relation ville – port en Chine (Port governance and port city interactions in China). *Les Cahiers Scientifiques du Transport*, 44: 25–54.

Wang. J.J. and Olivier. D. 2006. Port-FEZ bundles as spaces of global articulation: the case of Tianjin, China. *Environment & Planning A*, 38(8): 1487–1504.

Wang. J.J. and Olivier. D. 2007. "Chinese port cities in global supply chains". in Wang. J.J. et al. (Eds). 2007. *Ports, Cities, and Global Supply Chains*. Aldershot: Ashgate: 173–86.

Wang. J.J. and Olivier. D. 2007. "Hong Kong and Shenzhen: The Nexus in South China", in Cullinane, K., Song, D.-W. (Eds). *Asian Container Ports*. Houndmills, Basingstoke, UK: Palgrave MacMillan: 198–212.

Wang. J.J. and Slack. B. 2000. The evolution of a regional container port system: the Pearl River Delta. *Journal of Transport Geography*, 8(4): 263–75.

Weber, M., 1921. Wirtschaft und Gesellschaft. Teilband 5: Die Stadt. In: Nippel, W. (Ed.), Max Weber Gesamtausgabe. Mohr Siebeck, Tübingen (Reprinted 1999).

World Bank. 2003. World Bank Reform Toolkit: module 3 Alternative Port Management Structures and Ownership Models. Washington DC: World Bank.

World Bank. 2009. Tracks from the past, transport for the future: China's Railway Industry 1990–2008 and its future plans and possibilities, available at https://openknowledge.worldbank.org/handle/10986/3197.

Wu, Chuanjun and Xiaozhen, Gao. 1989. Growth Pattern of Port Cities. Geographical Research, 8(4): 9–150 (Chinese).

Wu, Songdi (ed.). 2006. Puzzle of China's Hundreds of Years Economy: Port Cities and their Hinterland and China's Modernization. Jinan: Shandong Pictorial Publishing House (Chinese).

Wu, Songdi. 2004. The Spatial Process of Port – Hinterland and China's Modernization. Hebei Academic Journal, 24(3): 160–66 (Chinese).

Wu, Yuwen, Dexun, Peng. 1995. Guangzhou Port – – Accelerator of Guangzhou Metropolis Building. Economic Geography, 15(1): 85–92 (Chinese).

Xu, Jiang and Wang, James. (2012) "Reassembling the state in urban China", Asia Pacific Viewpoint, Vol. 53: 7–22.

Yao, Yongchao. 2005. The Spatial Movement of Japanese and Russian Economic Power in Northeast China in 1906 – 1931 – from the Perspectives of Port, Railway and the Changes of Goods Distribution Scope. Journal of Chinese Historical Geography, 20(1): 35–41 (Chinese).

Zhao, Huangting. 2006. Fanyu Was the Earliest Beginning Port for Marine Silk Road in South China. Geography Science, 26(1): 118–27 (Chinese).

Zheng, Tianxiang. 2005. Competition and Cooperation of the Port Group in Greater Pearl River Delta. Port Economy, 3: 27–8 (Chinese).

Zuo, Zheng. 2002. On the Historical Devolvement of Guangzhou Port and Foreign Trade, Journal of Jinan University (Philosophy & Social Science Edition), 24(5): 39–46 (Chinese).

Index

Milton Keynes UK
Ingram Content Group UK Ltd.
UKHW031150141024
449569UK00024B/929